基于ArcGIS Engine 地理信息系统开发技术与实践

李进强　编著

武汉大学出版社

图书在版编目(CIP)数据

基于 ArcGIS Engine 地理信息系统开发技术与实践/李进强编著 .—武汉:武汉大学出版社,2017.11(2019.1 重印)

ISBN 978-7-307-16611-0

Ⅰ.基… Ⅱ.李… Ⅲ. 地理信息系统—系统开发 Ⅳ.P208

中国版本图书馆 CIP 数据核字(2017)第 259513 号

责任编辑:鲍 玲　　责任校对:李孟潇　　版式设计:汪冰滢

出版发行:**武汉大学出版社**　(430072 武昌 珞珈山)

(电子邮件:cbs22@ whu.edu.cn 网址:www.wdp.com.cn)

印刷:湖北金海印务有限公司

开本:787×1092 1/16 印张:22. 25 字数:525 千字 插页:1

版次:2017 年 11 月第 1 版 2019 年 1 月第 2 次印刷

ISBN 978-7-307-16611-0 定价:49. 80 元

前　言

ArcGIS Engine 是 Esri 公司推出的用于构建定制 GIS 应用程序的组件库,功能强大,目前已经成为 GIS 二次开发的主流工具。

本书将 GIS 系统中常见功能分为若干专题,按软件工程思想,从功能设计、详细设计、功能实现、系统集成等过程对每个专题进行了相应描述。对每部分所涉及的 ArcGIS Engine 接口、实现接口的类,以及对应的属性和方法进行了相应介绍。作者通过对源代码反复优化,提炼清晰思路和操作步骤,引入第三方控件使程序界面简化又符合潮流。每个功能按实用化要求实现,力求使读者既深入理解 GIS 系统开发原理与方法,而且能有效掌握 ArcGIS Engine 开发的实战技术。

全书共 23 章,第 1、第 2 章介绍 ArcGIS Engine 应用程序框架,包括系统界面设计与实现,视图切换等;第 2~第 8 章介绍地图符号化和专题制图,包括简单符号配置、图层标注、唯一值渲染、分级渲染、统计图表渲染、栅格数据渲染等;第 9~第 10 章介绍空间查询,包括属性查询、空间关系查询;第 11~第 16 章介绍矢量数据和栅格数据空间分析、包括缓冲区分析、矢量数据叠加分析、栅格数据重分类、成本距离分析、运输网络分析、几何网络分析等;第 17~第 20 章介绍空间数据库管理,包括数据访问,拓扑检查,数据建库,属性表展示等;第 21~第 22 章介绍三维数据展示与分析,包括三维数据展示、TIN 生成、表面分析等。全部基于 ArcGIS Engine 10. 5+Visual Studio2015. NET 开发环境(目前最新版本),C#开发语言进行讲解,附有相关专题的源代码实例,以方便读者理解和练习。

本书是作者多年使用 ArcGIS Engine 进行教学和项目研发的经验总结,可作为地理信息科学专业、测绘工程、遥感科学与技术、资源环境类专业或计算机相关专业的教材或实践指导书,也可作为从事地理信息系统开发的技术人员的参考资料。

由于作者水平有限,编写时间仓促,书中错漏之处在所难免,敬请读者批评指正。作者邮箱:361639771@ qq. com。

目　　录

第1章　ArcGIS Engine 应用程序框架

1.1　概述

ArcGIS Engine 提供了一个公共的开发控件集合，这些控件通过简单的绑定就可协同工作（如 TOCControl 可绑定 MapControl，它们称为伙伴控件），和其他第三方控件（组件）结合可以创建高度定制化的应用程序。

ArcGIS Engine 提供的开发控件有：

- ◆ 地图控件（MapControl）；
- ◆ 页面布局控件（PageLayoutControl）；
- ◆ 内容列表控件（TOCControl）；
- ◆ 工具条控件（ToolbarControl）；
- ◆ 场景控件（SceneControl）；
- ◆ 球体控件（GlobeControl）；
- ◆ 阅读者控件（ReaderControl）；
- ◆ 使用工具条控件的命令与工具集合。

使用 ArcGIS Engine 模板可以无编码搭建一个类似 ArcMap 的 GIS 程序框架，这种方式可以帮助初级用户在较短的时间内掌握技术要领，降低了开发者的门槛。但由于这个框架不支持窗口停靠，界面风格也不够流行等问题，对于实用化的开发是不够的。

1.2　功能描述

本章建议在 VS 环境下，使用 MapControl、PageLayoutControl、TOCControl 结合第三方控件实现较为流行的 GIS 框架，共包括两个第三方控件，实现界面如图 1-1 所示：

①WeifenLuo. WinFormsUI. Docking 控件，实现窗体停靠；

②DotNetBar 控件，实现 Ribbon 风格。

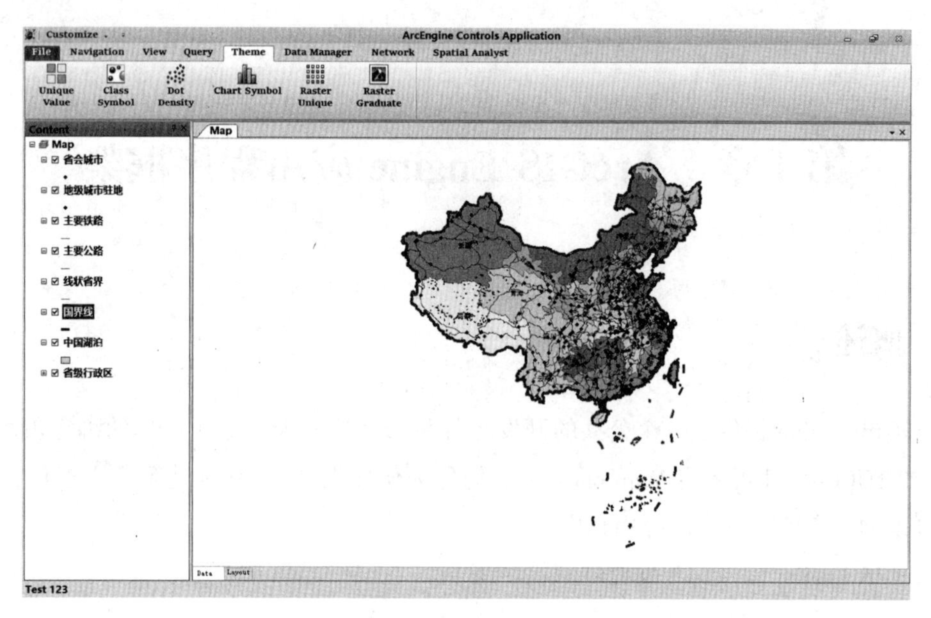

图 1-1　实现界面

1.3　系统实现

1.3.1　创建 ArcEngine 应用程序底稿

使用 ArcEngine 模板创建应用程序底稿比较方便，除自动建立 Windows 程序基本元素，还设置好了 ArcEngine 需要的版本绑定代码等。

启动 VS2015，选择"文件"→"新建"→"项目"，在项目类型中选择：

"Visual C#"→"ArcGIS"→"Extending ArcObjects"分类目录，再选择 MapControl Application 模板。指定项目存放位置(如 C：\ 用户目录)，输入项目名称(默认 MapControlApplicaton1)，点击"确定"，建立模板化的应用程序框架。

1.3.2　主窗体设计

1. 界面设计

①MainForm 窗体上保留 StatusStrip，AxLicenseControl 控件，删除 AxMapControl，AxTOCControl，AxToolBarControl，MenuStrip。

设置 MainForm 为无边框类型，即 FormBorderStyle=none；

设置 MainForm 为 MDI 窗体，即 IsMdiContainer=true。

②DockPanel/DotNetBar 控件添加到工具箱。

将 weiFenLuo. winFormsUI. Docking. dll、DevComponents. DotNetBar2. dl 拷贝到 bin\debug 目录，在工具箱添加两个选项卡命名为 WeiFenLuo，DotNetbar，右键点击该选项卡，出现

浮动菜单，以 WeiFenLuo 为例，之后操作步骤为：

"选择项"→"浏览"→"weiFenLuo. winFormsUI. Docking. dll"→"确定"。

此时工具箱出现 DockPanel 控件。

③将 DotNetbar 选项卡 RibbonControl 控件拖入 MainForm 窗体，设置 Dock = Top；此部分为 Ribbon 风格的菜单条代替传统菜单和工具条。将 WeiFenLuo 选项卡 DockPanel 控件拖入 MainForm 窗体空白处，设置 Dock = Fill；此部分为其他窗体停靠区域。

④添加导航 Ribbon 工具。

右键点击 RibbonControl1 眉头，再点击【Create Ribbon Tab】菜单项，创建一个 RibbonTabItem(实际上已有一个)，设置 Text 属性为 Navigation；

右键点击 Navigation 空白处，点击【Create RibbonBar】菜单项，创建一个 RibbonBar(默认名为：RibbonBar1)；

右键点击 RibbonBar1，点击【Add Button】菜单项，添加 5 个 ButtonItem，分别命名为 btnAddData，BtnPan，BtnZoomOut，btnZoomIn，BtnFullExtent。

包含关系及含义见表 1-1：

表 1-1　**包含关系及含义**

RibbonControl	RibbonTabItem	RibbonBar	ButtonItem	含义
RibbonControl1	Navigation	RibbonBar1	btnAddData	添加数据
			BtnPan	移屏
			BtnZoomOut	放大
			btnZoomIn	缩小
			BtnFullExtent	全图

分别双击 5 个 ButtonItem 建立 Click 事件响应函数，在其中分别激活 ArcEngine 相应命令类。

2. 功能实现

功能实现代码如下：

```
public sealed partial class MainForm:Form
{
    //私有成员：
    private string m_mapDocumentName=string.Empty;
    private TOCControlDockFrm m_frmTOCControl=null;
    private MapControlDockFrm m_frmMapControl=null;
    //定义 AxMapControl 属性,方便将来使用
    private AxMapControl _AxMapControl
    {
        get { return m_frmMapControl._AxMapControl; }
```

```
        }
        //定义 AxTOCControl 属性,方便将来使用
        private AxTOCControl _AxTOCControl
        {
            get { return m_frmTOCControl._AxTOCControl;}
        }
        //定义 IMapControl3 属性,方便将来使用
        private IMapControl3 _mapControl
        {
            get {
                IMapControl3 mapControl = (IMapControl3)(m_frmMapControl._
AxMapControl.Object);
                return mapControl;
                }
        }
        //构造函数
        public MainForm()
        {
            InitializeComponent();
        }
        //装载时间响应函数
        private void MainForm_Load(object sender,EventArgs e)
        {
            //初始化子两个窗体 Map,TOC,子窗体下节介绍
            m_frmTOCControl = new TOCControlDockFrm();
            m_frmMapControl = new MapControlDockFrm();

            //设置绑定
            m_frmTOCControl.SetBudderControl(m_frmMapControl._AxMap-
Control);
            m_frmMapControl.SetStatusLabel(this.statusBarXY);

            //DockState 为窗体的停靠状态
            m_frmTOCControl.Show(dockPanel1,DockState.DockLeft);
            m_frmMapControl.Show(dockPanel1,DockState.Document);
        }

        //Navigation 工具响应函数
```

```
    private void btnAddData_Click(object sender,EventArgs e)
    {
        ICommand pCommand=new ControlsAddDataCommand();
        pCommand.OnCreate(_mapControl.Object);
        pCommand.OnClick();
    }

    private void btnPan_Click(object sender,EventArgs e)
    {
        ICommand pCommand=new ControlsMapPanTool();
        pCommand.OnCreate(_AxMapControl.Object);
        _AxMapControl.CurrentTool=pCommand as ITool;
        pCommand=null;
    }

    private void btnZoomOut_Click(object sender,EventArgs e)
    {
        ICommand pCommand=new ControlsMapZoomOutTool();
        pCommand.OnCreate(_AxMapControl.Object);
        _AxMapControl.CurrentTool=pCommand as ITool;
        pCommand=null;
    }

    private void btnZoomIn_Click(object sender,EventArgs e)
    {
        ICommand pCommand=new ControlsMapZoomInTool();
        pCommand.OnCreate(_AxMapControl.Object);
        _AxMapControl.CurrentTool=pCommand as ITool;
        pCommand=null;
    }

    private void btnFullExtent_Click(object sender,EventArgs e)
    {
        ICommand pCommand=new ControlsMapFullExtentCommand();
        pCommand.OnCreate(_mapControl.Object);
        pCommand.OnClick();
    }
}
```

代码中 MapControlDockFrm，TOCControlDockFrm 两个类分别是地图(Map)窗体和内容列表(Table Of Contents)窗体，将在后续章节介绍，编译时可先隐去与之相关的代码。

1.3.3 MapControlDockFrm 窗体设计

1. 界面设计

①新建 Windows 窗体，取名为 MapControlDockFrm。

②添加 TabControl 控件：包括两个属性页：

将 TabControl 控件拖入 MapControlDockFrm，Dock 属性设置为 Fill。将 Alignment 属性设置为 Bottom。点击 TabPages 属性右边的按钮，弹出 TabPage 集合编辑器：

a. 添加属性页 tabPage1，Text 设置为“Data”；

b. 添加属性页 tabPage2，Text 设置为“Layout”。

③添加 GIS 图形控件：

a. 选择“Data”选项卡，拖入 AxMapControl 控件，设置 Dock 属性为 Fill；

b. 选择“Layout”选项卡，拖入 AxPageLayoutControl 控件，设置 Dock 属性为 Fill。

2. 功能实现

最主要修改 MapControlDockFrm 基类 Form 为 DockContent，这是关键一步，DockContent 属于 weiFenLuo. winFormsUI. Docking. dll 链接库，该类支持在 DockPanel 控件定义的区域上的停靠特性，实现代码如下：

```
public partial class MapControlDockFrm:DockContent
{
    //私用成员
    //两个常用接口 IMapControl3/IPageLayoutControl2：
    private IMapControl3 m_mapControl=null;
    private IPageLayoutControl2 m_pageLayoutControl=null;
    //显示鼠标所处的 XY 坐标：
    private ToolStripStatusLabel m_statusBarXY=null;

    //暴露 AxMapControl 对象,方便其他控件绑定
    public AxMapControl _AxMapControl
    {
        get  {  return this.axMapControl1;  }
    }
    //暴露 AxPageLayoutControl 对象,方便其他控件绑定
    public AxPageLayoutControl _AxPageLayoutControl
    {
        get {  return this.axPageLayoutControl1; }
    }
    //提供设置 StatusLabel 的公用方法,实现向父窗体发送坐标位置信息
```

```
        public void SetStatusLabel(ToolStripStatusLabel statusBarXY)
        {
            m_statusBarXY = statusBarXY;
        }

        //构造函数
        public MapControlDockFrm()
        {
            InitializeComponent();
        }

        //装载事件响应函数
        private void MapControlDockFrm_Load(object sender,EventArgs e)
        {
            m_mapControl = (IMapControl3)(_AxMapControl.Object);
            m_pageLayoutControl = ( IPageLayoutControl2 ) _AxPageLayout-
Control.Object;
        }

        //鼠标移动响应函数,对 m_statusBarXY 的 Text 属性赋值
        private void axMapControl1_OnMouseMove(object sender,IMapCon-
trolEvents2_OnMouseMoveEvent e)
        {
            m_statusBarXY.Text = string.Format("{0},{1}  {2}",
        e.mapX.ToString("#######.##"),
        e.mapY.ToString("#######.##"),
        axMapControl1.MapUnitsTo String ().Substring(4));
        }
    }
```

1.3.4　TOCControlDockFrm 窗体设计

1. 界面设计

①新建 Windows 窗体，取名为 TOCControlDockFrm。

②添加 GIS 图形控件，拖入 AxTOCControl 控件，设置 Dock 属性为 Fill。

③添加图层操作浮动菜单，拖入 ContextMenuStrip，命名为 contextMenuTOCLyr，添加菜单项 RemoveLayer、ZoomToLayer、Symbolize、OpenAttributeTable 等，有些供以后使用。

2. 类设计

首先修改 TOCControlDockFrm 基类 Form 为 DockContent，实现代码如下：

```
public partial class TOCControlDockFrm:DockContent
{
    public TOCControlDockFrm()
    {
        InitializeComponent();
    }
    //暴露 AxTOCControl 对象,方便其他控件绑定
    public AxTOCControl _AxTOCControl
    {
        get { return this.axTOCControl1; }
    }
    //设置私有 IMapControl3 对象,方便其他方法访问 Map
    private IMapControl3 _mapControl
    {
        get {
            IMapControl3 mapControl = this.axTOCControl1.Buddy as
IMapControl3;
            return mapControl;
        }
    }
    //绑定 AxMapControl 控件
    public void SetBudderControl(AxMapControl _AxMapControl)
    {
        this.axTOCControl1.SetBuddyControl(_AxMapControl);
    }
    //右键响应函数
    private void axTOCControl1_OnMouseDown(object sender,
    ITOCControlEvents_OnMouse DownEvent e)
    //移除图层
    private void removeLayerToolStripMenuItem_Click(object sender,
EventArgs e)
    //缩放至图层
    private void zoomToLayerToolStripMenuItem_Click(object sender,
EventArgs e)
}
```

3. 功能实现

鼠标右键响应函数的主要功能是：

①记录鼠标选中的图层和图例对象，为接下来的图层操作提供输入参数；

②展开图层操作浮动菜单：

首先用 TOCControl 的 HitTest 函数测试鼠标点击的图层接口对象，然后用私有成员变量分别记录图层和图例对象，代码如下：

```
    private ILayer m_tocRightLayer=null;
    private ILegendClass m_tocRightLegend=null;
    private void axTOCControl1_OnMouseDown(object sender,ITOCControlEv-
ents_OnMouseDownEvent e)
    {
        if (e.button!= 2)
            return;
        esriTOCControlItem itemType = esriTOCControlItem.esriTOCControlI-
temNone;
        IBasicMap basicMap=null;
        ILayer layer=null;
        object unk=null;
        object data=null;
        //this.axTOCControl.GetSelectedItem(ref itemType,ref basicMap,
ref layer,ref unk,ref data);
        this.axTOCControl1.HitTest(e.x,e.y,ref itemType,ref basicMap,
ref layer,ref unk,ref data);
        switch (itemType)
        {
            case esriTOCControlItem.esriTOCControlItemLayer:
                this.m_tocRightLayer=layer;
                this.m_tocRightLegend=null;
                this.contextMenuTOCLyr.Show(this.axTOCControl1,e.x,e.y);
                break;
            case esriTOCControlItem.esriTOCControlItemLegendClass:
                this.m_tocRightLayer=layer;
                this.m_tocRightLegend = ((ILegendGroup)unk).get_Class
((int)data);
                this.contextMenuTOCLyr.Show(this.axTOCControl1,e.x,e.y);
                break;
            case esriTOCControlItem.esriTOCControlItemMap:
                //this.contextMenuTOCMap.Show(this.axTOCControl1,e.x,e.y);
                break;
        }
    }
```

```
    //移除图层
    private void removeLayerToolStripMenuItem_Click(object sender, EventArgs e)
    {
        _mapControl.Map.DeleteLayer(m_tocRightLayer);
    }
    //缩放至图层
    private void zoomToLayerToolStripMenuItem_Click(object sender, EventArgs e)
    {
        _mapControl.Extent=m_tocRightLayer.AreaOfInterest;
    }
```

1.4 License 配置

配置 License 简单方法是通过设置 LicenseControl，操作方法是：右键点击 LicenseControl，点击属性菜单。浏览弹出的对话框，其中 ArcGIS Engine 已经选中，如果需要其他扩展模块的许可，可以在右侧选中对应的复选框，点击"确定"按钮，如图 1-2 所示。

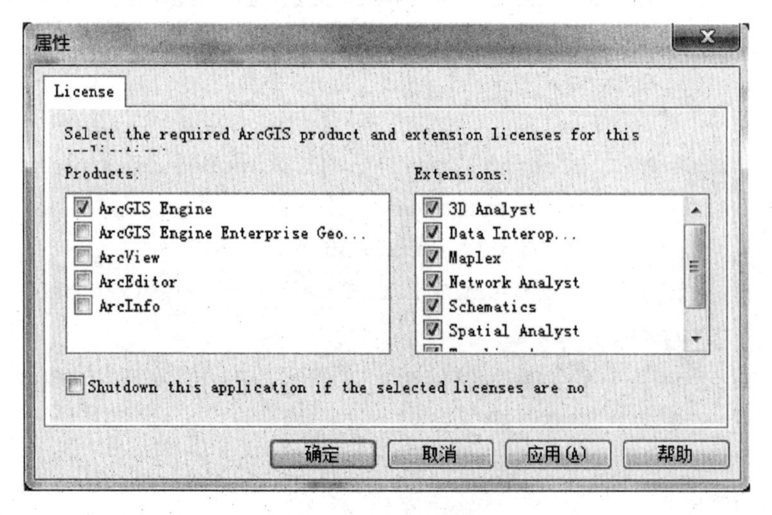

图 1-2 属性对话框

LicenseControl 配置有可能出现有些功能授权不正常的现象。ESRI 推荐在运行时配置 License，也是本文推荐的做法。

具体做法是修改 Program 类的 Main( ) 函数，代码如下：

```
    /// <summary>
    /// The main entry point for the application.
```

```
    /// </summary>
    [STAThread]
    static void Main()
    {
        //版本绑定
        if (!RuntimeManager.Bind(ProductCode.Engine))
        {
            if(!RuntimeManager.Bind(ProductCode.Desktop))
            {
                MessageBox.Show("Unable to bind to ArcGIS runtime. Ap-
plication will be shut down.");
                return;
            }
        }

        //AoLicense 初始化
        LicenseInitializer aoLicenseInitializer=new LicenseInitializer();
        if (!aoLicenseInitializer.InitializeApplication (new esriLicense
ProductCode[]
                                    { esriLicenseProductCode.esriLicensePro-
ductCodeAdvanced, esriLicenseProductCode.esriLicensePro-ductCodeEngi-
neGeoDB },
                                     new esriLicenseExtensionCode[]
                                    { esriLicenseExtensionCode.esriLicenseEx-
tensionCodeDataInteroperability,esriLicenseExtensionCode. esriLicense-
ExtensionCodeSpatialAnalyst, esriLicenseExtensionCode.esriLicense-Ex-
tensionCode3DAnalyst, esriLicenseExtensionCode.esriLicense-Extension-
CodeNetwork}))
        {
            System.Windows.Forms.MessageBox.Show ( " This application
could not initialize
                                    with the correct ArcGIS license and will
shutdown. LicenseMessage:" +
                                    aoLicenseInitializer.LicenseMessage());
            aoLicenseInitializer.ShutdownApplication();
            return;
        }
```

```
        Application.EnableVisualStyles();
        Application.SetCompatibleTextRenderingDefault(false);
        Application.Run(new MainForm());

        aoLicenseInitializer.ShutdownApplication();
        ESRI.ArcGIS.ADF.COMSupport.AOUninitialize.Shutdown(); //释放 Com 资源
    }
```

这里用到 LicenseInitializer 密封类，Ersi 给出详细的实现代码，详见附录 3。

1.5 编译运行

按 F5 键即可编译运行程序，至此桌面 GIS 应用程序框架已经搭建好了，可以通过 Ribbon 的工具打开地图数据，浏览地图。

第2章　MapControl与PageLayoutControl同步

2.1　知识要点

在ArcMap中，能够很方便地进行MapView和Layout View两种视图的切换，而且二者之间的数据是同步显示的。ArcEngine实现两种视图同步有多种方法，本章介绍的方法是通过PageLayoutControl和MapControl中的Map对象指向同一个地图实例(二者共享一份地图)，要点如下：

①使用PageLayoutControl的ReplaceMaps()方法，MapControl的Map属性，指向MapDocument中的同一个Map对象。

②为保证系统中只有一个ActiveView对象，当从PageLayoutControl切换至MapControl时，需要先调用PageLayoutControl的DeActivate方法，再调用MapControl的Activate方法；从MapControl切换至PageLayoutControl时，先调用MapControl的DeActivate方法，再调用PageLayoutControl的Activate方法。避免地图出现闪烁现象。

③ReplaceMaps()方法只接收IMaps类型的对象(IMap集合)，AE中没有IMaps的实现类，这需要由自己动手实现该接口。

2.2　新建同步控制类ControlsSynchronizer

1. 类设计

在解决方案面板中右击项目名，选择“添加”→“类”，在类别中选择“Visual C#项目项”，在模板中选择“类”，输入类名“ControlsSynchronizer. cs”，类设计如下：

```
public class ControlsSynchronizer
{
    //类成员
    private IMapControl3 m_mapControl=null;
    private IPageLayoutControl2 m_pageLayoutControl=null;
    private bool m_IsMapCtrlactive=true;

    //构造函数,为类成员赋值
     public ControlsSynchronizer(IMapControl3 mapControl,IPageLayout-
Control2 pageLayoutControl)
```

```
{
    m_mapControl=mapControl;
    m_pageLayoutControl=pageLayoutControl;
}

//=========属性============
//当前 ActiveView 的类型
public string ActiveViewType
{
    get
    {
        if (m_IsMapCtrlactive)
            return "MapControl";
        else
            return "PageLayoutControl";
    }
}
//当前活动的 Control
public object ActiveControl
{
    get
    {
        if (m_IsMapCtrlactive)
            return m_mapControl.Object;
        else
            return m_pageLayoutControl.Object;
    }
}

//============方法=============
public void ActivateMap()
public void ActivatePageLayout()
public void BindControls(bool activateMapFirst)
}
```

2. 类实现

①ActivateMap()函数：激活 MapControl 并解除 the PagleLayoutControl，具体代码如下：

```
public void ActivateMap()
{
```

```
        try
        {
            //解除 PagleLayout
            m_pageLayoutControl.ActiveView.Deactivate();
            //激活 MapControl
            m_mapControl.ActiveView.Activate(m_mapControl.hWnd);
            m_IsMapCtrlactive=true;
        }
        catch (Exception ex)
        {
            throw new Exception(string.Format("ControlsSynchronizer::
ActivateMap:\r\n{0}",ex.Message));
        }
    }
```

②ActivatePageLayout()函数：激活 PagleLayoutControl 并解除 MapCotrol，具体代码如下：

```
    public void ActivatePageLayout()
    {
        try
        {
        //解除 MapControl
            m_mapControl.ActiveView.Deactivate();
            //激活 PageLayoutControl
            m_pageLayoutControl.ActiveView.Activate(m_pageLayoutCon-
trol.hWnd);
            m_IsMapCtrlactive=false;
        }
        catch (Exception ex)
        {
            throw new Exception(string.Format("ControlsSynchronizer::
ActivatePageLayout:\r\n{0}",ex.Message));
        }
    }
```

③BindControls()函数：使 MapControl 和 PageLayoutControl 指定共同的 Map，具体代码如下：

```
    public void BindControls(bool activateMapFirst)
    {
        //创造 IMap,IMaps 实例,
```

```
        IMap newMap = new MapClass();
        newMap.Name = "Map";
        IMaps maps = new Maps();
        maps.Add(newMap);

        //调用 PageLayout 的 ReplaceMaps 来置换 focus map
        m_pageLayoutControl.PageLayout.ReplaceMaps(maps);
        //把 newMap 赋给 MapControl
        m_mapControl.Map = newMap;

        //确定最后活动的 control 被激活
        if (activateMapFirst)
            this.ActivateMap();
        else
            this.ActivatePageLayout();
    }
```

2.3 新建 Maps 类

在同步类中用到 Maps 类，该类实现 IMaps 接口相关功能，它是 IMap 集合用于管理地图对象。Maps 类代码比较简单，现列出详细代码如下：

```
public class Maps:IMaps,IDisposable
{
    private ArrayList m_array = null;
    public Maps()
    {
        m_array = new ArrayList();
    }

    public void Dispose()
    {
        if (m_array != null)
        {
            m_array.Clear();  m_array = null;
        }
    }

    public int Count
```

```
        {
            get {   return m_array.Count; }
        }

        public IMap Create()
        {
            IMap newMap = new MapClass();
            m_array.Add(newMap);
            return newMap;
        }
        public void RemoveAt(int Index)
        {
            if (Index > m_array.Count || Index < 0)
                throw new Exception("Maps::RemoveAt: \r\nIndex is out of
range!");
            m_array.RemoveAt(Index);
        }

        public void Reset()
        {
            m_array.Clear();
        }

        public IMap get_Item(int Index)
        {
            if (Index > m_array.Count || Index < 0)
                throw new Exception("Maps::get_Item: \r\nIndex is out of
range!");

            return m_array[Index] as IMap;
        }

        public void Remove(IMap Map)
        {
            m_array.Remove(Map);
        }

        public void Add(IMap Map)
```

```
        {
            if (Map== null)
                throw new Exception("Maps::Add: \r\nNew Map is mot ini-
tialized!");
            m_array.Add(Map);
        }
    }
```

2.4 两种视图的同步

①在 MapControlDockFrm 类中添加成员变量，即同步类对象：

```
private ControlsSynchronizer m_controlsSynchronizer=null;
private IMapControl3 m_mapControl=null;
private IPageLayoutControl2 m_pageLayoutControl=null;
private ToolStripStatusLabel m_statusBarXY=null;
```

修改 MapControlDockFrm_ Load 函数中进行初始化函数：

```
//get the MapControl/PageLayoutControl
m_mapControl=(IMapControl3)(_AxMapControl.Object);
m_pageLayoutControl = (IPageLayoutControl2)_AxPageLayoutControl.
Object;

//初始化 controls synchronization calss
m_controlsSynchronizer=new ControlsSynchronizer(m_mapControl,m_
pageLayoutControl);
//把 MapControl 和 PageLayoutControl 绑定起来(指向同一个 Map),并设置 MapControl
为活动的 Control
m_controlsSynchronizer.BindControls(true);
```

②建立 TabControl 的页切换响应函数：

```
private void tabControl1_SelectedIndexChanged(object sender,Even-
tArgs e)
{
    switch (this.tabControl1.SelectedIndex)
    {
        case 0:
            //激活 MapControl
            m_controlsSynchronizer.ActivateMap();
            break;
        case 1:
```

```
            //激活 PageLayoutControl
            m_controlsSynchronizer.ActivatePageLayout();
            break;
        }
    }
```

2.5 编译运行

按 F5 键，编译运行程序，点击 TabControl 页可切换 MapControl 和 PageLayoutControl 两种视图。

第3章　图层符号选择器的实现

3.1　知识要点

更改图层符号样式有两个途径：

1. 更改图层渲染器的符号属性

利用 IGeoFeatureLayer 接口获取图层渲染器 Renderer，然后更改 Renderer 的 Symbol（符号属性），即可更改要素符号样式，要素图层也就实现了 IGeoFeatureLayer 接口，基本步骤如下：

```
//1:创建渲染器,并设置新符号
ISimpleRenderer renderer=new SimpleRendererClass();
renderer.Symbol=pLineSymbol as ISymbol;
//2:为图层设置渲染器
IGeoFeatureLayer geoFeatureLyr=this.m_tocRightLayer as IGeoFeatu-
reLayer;
geoFeatureLyr.Renderer=renderer as IFeatureRenderer;
//3:更新 Map 控件和图层控件
this.axMapControl1.ActiveView.Refresh();
this.axTOCControl1.Update();
```

2. 更改图层图例的符号属性

利用 ILegendInfo 接口获得图层 ILegendGroup 对象（图例组），再从 LegendGroup 获得 ILegengd 对象（ILegendGroup 中每一个 Item 是 ILegend 对象（图例对象））。修改 ILegend 的符号属性，可更改要素符号样式。要素图层就实现了 ILegendInfo 接口。具体步骤如下：

```
//1:获取图例
ILegendInfo lgInfo=this.m_tocRightLayer as ILegendInfo;
ILegendGroup lgGroup=lgInfo.get_LegendGroup(0);
ILegendClass lgClass=lgGroup.get_Class(0);
//2:更改图例符号
lgClass.Symbol=pLineSymbol as ISymbol;
//3:更新主 Map 控件和图层控件
this.axMapControl1.ActiveView.Refresh();
this.axTOCControl1.Update();
```

不论用哪种方式，都必须先准备需要的符号，ArcGIS Engine 提供了 SymbologyControl 控件，极大地方便了图层符号的选择。本章介绍如何利用 SymbologyControl 控件选择符号改变图层符号样式。

3.2　功能描述

本章实现的符号选择器有如下功能：右键点击 TOCControl 控件中图层，在图层操作浮动菜单上点击 Symbolize 菜单项，弹出符号选择对话框，如图 3-1 所示，对话框能够根据图层几何类型自动加载相应的符号(如点、线、面)。用户可以调整符号的颜色、线宽、角度等参数，还可以打开自定义的符号文件(＊ServerStyle)加载更多的符号。

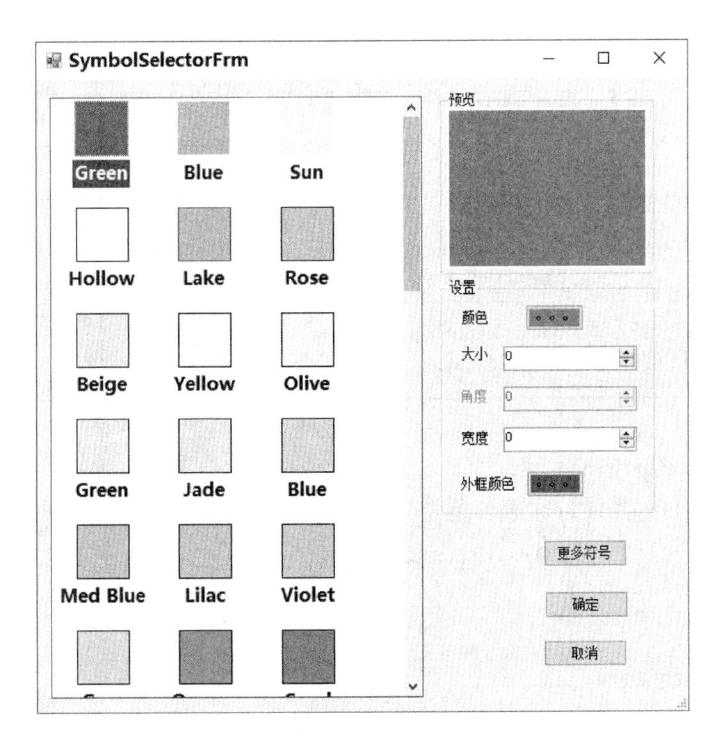

图 3-1

3.3　功能实现

1. 符号选择器界面设计

新建 Winodws 窗体，命名为 SymbolSelectorFrm，修改窗体的 Text 属性为“选择符号”。并添加 SymboloryControl、PictureBox、Button、NumericUpDown、ColorDialog、OpenFileDialog、ContextMenuStrip 控件，详见表 3-1(Label、GroupBox 未列在表中)。

表 3-1　　控件及其属性、含义

控　件	Name 属性	含　义	其　他
SymbologyControl	axSymbologyControl	Esri 符号控件	
PictureBox	ptbPreview	预览图片	
NumericUpDown	nudSize	大小	
NumericUpDown	nudWidth	线宽	
NumericUpDown	nudAngle	角度	
Button	btnColor	颜色	
Button	btnOutlineColor	外框颜色	
Button	btnMoreSymbols	更多符号	
Button	btnOK	确定	DialogResult 设为 OK
Button	btnCancel	取消	
ColorDialog	colorDialog	颜色对话框	
OpenFileDialog	openFileDialog	文件对话框	
ContextMenuStrip	contextMenuStrip MoreSymbol	浮动菜单	

2. SymbolSelectorFrm 实现

SymbolSelectorFrm 类设计代码如下：

```
public partial class SymbolSelectorFrm:Form
{
    //私有成员
    private ILayer m_pLayer;
    private ILegendClass m_pLegendClass;
    private IStyleGalleryItem m_pStyleGalleryItem=null;
    //暴露 Symbol 属性
    public ISymbol _Symbol
    {
        get
        {
            if (m_pStyleGalleryItem!= null)
                return (ISymbol)m_pStyleGalleryItem.Item;
            else
                return null;
```

```
        }
    }

    //构造函数,初始化全局变量
    public SymbolSelectorFrm(ILegendClass tempLegendClass,ILayer tem-
pLayer)

    //加载事件响应函数
    private void SymbolSelectorFrm_Load(object sender,EventArgs e)
    //SymbologyControl 鼠标点击事件,符号预览
    private void axSymbologyControl_OnMouseDown(object sender,ISym-
bologyControlEvents_OnMouseDownEvent e)
    //选中符号时触发的事件
    private void axSymbologyControl_OnItemSelected(object sender,ESRI.
ArcGIS.Controls.ISymbologyControlEvents_OnItemSelectedEvent e)
    //参数调节响应函数
    private void nudSize_ValueChanged(object sender,EventArgs e)
    private void nudAngle_ValueChanged(object sender,EventArgs e)
    private void nudWidth_ValueChanged(object sender,EventArgs e)
    private void btnColor_Click(object sender,EventArgs e)
    private void btnOutlineColor_Click(object sender,EventArgs e)
    private void btnMoreSymbols_Click(object sender,EventArgs e)
    //确定取消
    private void btnOK_Click(object sender,EventArgs e)
    private void btnCancel_Click(object sender,EventArgs e)

    //辅助函数
    private void SetFeatureClassStyle(esriSymbologyStyleClass symbolo-
gyStyleClass)
    public Color IRgbColorToColor(IRgbColor pRgbColor)
    public IColor ColorToIRgbColor(Color color)
}
```

①修改 SymbolSelectorFrm 的构造函数，传入图层和图例接口。代码如下：

```
    //构造函数,初始化全局变量
    public SymbolSelectorFrm(ILegendClass tempLegendClass,ILayer tem-
pLayer)
    {
        InitializeComponent();
```

```
        this.m_pLegendClass=tempLegendClass;
        this.m_pLayer=tempLayer;
    }
```

②SymbolSelectorFrm_ Load 事件响应函数：根据图层类型(如点、线、面)为 SymbologyControl 导入相应的符号样式文件，并设置参数调节控件的可视性。这里要用到辅助函数 SetFeatureClassStyle()，代码如下：

```
    private void SymbolSelectorFrm_Load(object sender,EventArgs e)
    {
        //取得 ArcGIS 安装路径
        string sInstall=ESRI.ArcGIS.RuntimeManager.ActiveRuntime.Path;
        //载入 ESRI.ServerStyle 文件到 SymbologyControl
        this.axSymbologyControl.LoadStyleFile(sInstall+"Styles\\ESRI.
ServerStyle");
        //依据图层的类型(点线面);设置 SymbologyControl 的 StyleClass;设置各控件的可
用性
        switch(((IFeatureLayer)m_pLayer).FeatureClass.ShapeType)
        {
            case ESRI.ArcGIS.Geometry.esriGeometryType.esriGeometry-
Point:
                this.SetFeatureClassStyle(esriSymbologyStyleClass.es-
riStyleClassMarkerSymbols);
                SetControlValidated(00001111);
                break;
            case ESRI.ArcGIS.Geometry.esriGeometryType.esriGeometry-
Polyline:
                this.SetFeatureClassStyle(esriSymbologyStyleClass.es-
riStyleClassLineSymbols);
                SetControlValidated(00110000);
                break;
            case ESRI.ArcGIS.Geometry.esriGeometryType.esriGeometry-
Polygon:
                this.SetFeatureClassStyle(esriSymbologyStyleClass.es-
riStyleClassFillSymbols);
                SetControlValidated(11110000);
                break;
            case ESRI.ArcGIS.Geometry.esriGeometryType.esriGeometry-
MultiPatch:
                this.SetFeatureClassStyle(esriSymbologyStyleClass.es-
```

```
riStyleClassFillSymbols);
                SetControlValidated(11110000);
                break;
            default:
                this.Close();
                break;
        }
    }

    private void SetControlValidated(int Flag)
    {
        this.lblAngle.Enabled=(Flag & 000000001) > 0;
        this.nudAngle.Enabled=(Flag & 000000010) > 0;
        this.lblSize.Enabled=(Flag & 000000100) > 0;
        this.nudSize.Enabled=(Flag & 000001000) > 0;
        this.lblWidth.Enabled=(Flag & 000100000) > 0;
        this.nudWidth.Enabled=(Flag & 001000000) > 0;
        this.lblOutlineColor.Enabled=(Flag & 010000000) > 0;
        this.btnOutlineColor.Enabled=(Flag & 100000000) > 0;
    }
```

③符号预览，当用户选定某一符号时，符号可以显示在 PictureBox 控件中，代码如下：

```
    ///把选中并设置好的符号在 picturebox 控件中预览
    private void axSymbologyControl_OnMouseDown(object sender,
                                                    ISymbologyControlEvents_
OnMouseDownEvent e)
    {
        if (e.button== 1)
        {
            m_pStyleGalleryItem=axSymbologyControl.HitTest(e.x,e.y);
            PreviewImage();
        }
    }

    //使用 ISymbologyStyleClass 接口的 PreviewItem 将符号定义画成 Image
    private void PreviewImage( )
    {
        ISymbologyStyleClass pStyleClass=null;
        pStyleClass= this. axSymbologyControl. GetStyleClass ( this. ax-
```

```
SymbologyControl.StyleClass);
        stdole.IPictureDisp picture=pStyleClass.PreviewItem(m_pStyle-
GalleryItem,this.ptbPreview.Width,this.ptbPreview.Height);
        Image image=System.Drawing.Image.FromHbitmap(new System.IntPtr
(picture.Handle));
        this.ptbPreview.Image=image;
    }
```

④双击“确定”按钮，添加如下代码：

```
    ///确定按钮
    private void btnOK_Click(object sender,EventArgs e)
    {
        //关闭窗体
        this.Close();
    }
```

⑤辅助函数，代码如下：

```
    ///SetFeatureClassStyle(),根据图层类型过滤符号库中的符号
    private void SetFeatureClassStyle(esriSymbologyStyleClass symbolo-
gyStyleClass)
    {
        this.axSymbologyControl.StyleClass=symbologyStyleClass;
        ISymbologyStyleClass pSymbologyStyleClass=
                                        this.axSymbologyControl.GetStyle-
Class(symbologyStyleClass);
        if (this.m_pLegendClass!= null)
        {
                IStyleGalleryItem currentStyleGalleryItem = new Server-
StyleGalleryItem();
            currentStyleGalleryItem.Name="当前符号";
            currentStyleGalleryItem.Item=m_pLegendClass.Symbol;
            pSymbologyStyleClass.AddItem(currentStyleGalleryItem,0);
            this.m_pStyleGalleryItem=currentStyleGalleryItem;
        }
        pSymbologyStyleClass.SelectItem(0);
    }

    /// 将 ArcGIS Engine 中的 IRgbColor 接口转换至.NET 中的 Color 结构
    public Color IRgbColorToColor(IRgbColor pRgbColor)
    {
```

```
        return ColorTranslator.FromOle(pRgbColor.RGB);
    }

    ///将.NET 中的 Color 结构转换至 ArcGIS Engine 中的 IColor 接口
    public IColor ColorToIRgbColor(Color color)
    {
        IColor pColor=new RgbColorClass();
        pColor.RGB=color.B*65536+color.G*256+color.R;
        return pColor;
    }
```

3.4 调用自定义符号选择器

通过以上操作，符号选择器雏形已经完成，在图层操作浮动菜单上添加菜单项——Symbolize，建立该菜单响应函数，具体代码如下：

```
    private void symbolizeToolStripMenuItem_Click(object sender,EventArgs e)
    {
        //取得图例
        ILegendClass pLegendClass=this.m_tocRightLegend;
        if (this.m_tocRightLegend== null)
        {
            ILegendInfo lgInfo=this.m_tocRightLayer as ILegendInfo;
            ILegendGroup lgGroup=lgInfo.get_LegendGroup(0);
            pLegendClass=lgGroup.get_Class(0);
        }

        //创建符号选择器 SymbolSelector 实例
        SymbolSelectorFrm Frm = new SymbolSelectorFrm ( pLegendClass,
this.m_tocRightLayer);
        if (Frm.ShowDialog()== DialogResult.OK)
        {
            //局部更新主 Map 控件
            _mapControl.ActiveView.PartialRefresh(esriViewDrawPhase.esriViewGeography,null,null);
            //设置新的符号
            pLegendClass.Symbol=Frm._Symbol;
            //更新主 Map 控件和图层控件
```

```
        this._mapControl.ActiveView.Refresh();
        this.axTOCControl1.Refresh();
    }
}
```

按 F5 键编译运行，比较简陋的符号选择器已经完成。

3.5 功能增强

3.5.1 符号参数调整

在地图整饰中，符号参数的调整是必需的功能。下面我们将实现符号基本颜色、多边形外框颜色、线符号线宽、点符号大小/角度等参数的调整。

(1)添加 SymbologyControl 的 OnItemSelected 事件

此事件在鼠标选中符号时触发，在事件响应函数中显示出选定符号的初始参数，代码如下：

```
///选中符号时触发的事件
private void axSymbologyControl_OnItemSelected(object sender,ESRI.ArcGIS.Controls.ISymbologyControlEvents_OnItemSelectedEvent e)
{
    m_pStyleGalleryItem=(IStyleGalleryItem)e.styleGalleryItem;
    IRgbColor pColor;
    switch (this.axSymbologyControl.StyleClass)
    {
        //点符号
        case esriSymbologyStyleClass.esriStyleClassMarkerSymbols:
            pColor=((IMarkerSymbol)m_pStyleGalleryItem.Item).Color as IRgbColor;
            //设置按钮背景色
            this.btnColor.BackColor=this.IRgbColorToColor(pColor);
            //设置点符号角度和大小初始值
            this.nudAngle.Value =(decimal)((IMarkerSymbol)m_pStyleGalleryItem.Item).Angle;
            this.nudSize.Value=(decimal)((IMarkerSymbol)m_pStyleGalleryItem.Item).Size;
            break;
        //线符号
        case esriSymbologyStyleClass.esriStyleClassLineSymbols:
            pColor=((ILineSymbol)m_pStyleGalleryItem.Item).Color
```

```
as IRgbColor;
                //设置按钮背景色
                this.btnColor.BackColor=this.IRgbColorToColor(pColor);
                //设置线宽初始值
                this.nudWidth.Value=(decimal)((ILineSymbol)m_pStyle-
GalleryItem.Item).Width;
                break;
            //面符号
            case esriSymbologyStyleClass.esriStyleClassFillSymbols:
                //设置按钮背景色
                pColor=((IFillSymbol)m_pStyleGalleryItem.Item).Color
as IRgbColor;
                this.btnColor.BackColor=this.IRgbColorToColor(pColor);
                //设置外框颜色
                pColor=((IFillSymbol)m_pStyleGalleryItem.Item).Outline.
Color as IRgbColor;
                this.btnOutlineColor.BackColor=this.IRgbColorToColor
(pColor);
                //设置外框线宽度初始值
                this.nudWidth.Value=(decimal)((IFillSymbol)m_pStyle-
GalleryItem.Item).Outline.Width;
                break;
            default:
                this.btnColor.BackColor=Color.Black;
                break;
        }

        //预览符号
        this.PreviewImage();
    }
```

(2)调整点符号的大小/角度

在 nudSize 控件的 ValueChanged 事件响应函数设置点符号的大小。在 nudAngle 控件的 ValueChanged 事件响应函数设置点符号的角度。具体代码如下：

```
/// 调整符号大小-点符号
    private void nudSize_ValueChanged(object sender,EventArgs e)
    {
        ((IMarkerSymbol)this.m_pStyleGalleryItem.Item).Size=(double)
this.nudSize.Value;
```

```
        this.PreviewImage();
    }

    ///调整符号角度-点符号
    private void nudAngle_ValueChanged(object sender,EventArgs e)
    {
        ((IMarkerSymbol)this.m_pStyleGalleryItem.Item).Angle=(double)
this.nudAngle.Value;
        this.PreviewImage();
    }
```

(3)调整线符号和面符号的线宽

nudWidth 控件的 ValueChanged 事件响应函数，设置线符号的线宽和面符号的外框线的线宽。代码如下：

```
    ///调整符号宽度-限于线符号和面符号
    private void nudWidth_ValueChanged(object sender,EventArgs e)
    {
        switch (this.axSymbologyControl.StyleClass)
        {
            case esriSymbologyStyleClass.esriStyleClassLineSymbols:
                ILineSymbol pLineSymbol=(ILineSymbol)this.m_pStyleGal-
leryItem.Item;
                pLineSymbol.Width=Convert.ToDouble(this.nudWidth.Value);
                break;
            case esriSymbologyStyleClass.esriStyleClassFillSymbols:
                //取得面符号的轮廓线符号
                ILineSymbol lSymbol=((IFillSymbol)this.m_pStyleGalle-
ryItem.Item).Outline;
                lSymbol.Width=Convert.ToDouble(this.nudWidth.Value);
                ((IFillSymbol)this.m_pStyleGalleryItem.Item).Outline=
lSymbol;
                break;
        }
        this.PreviewImage();
    }
```

(4)调整符号颜色

此处调用.NET 的颜色对话框 ColorDialog 选定颜色，修改颜色按钮的背景色为选定的颜色，同时修改选定样式中符号的颜色。btnColor/btnOutlineColor 按钮响应函数代码如下：

```
    private void btnColor_Click(object sender,EventArgs e)
```

```
    {
        //调用系统颜色对话框
        if (this.colorDialog.ShowDialog()= = DialogResult.OK)
        {
            //将颜色按钮的背景颜色设置为用户选定的颜色
            this.btnColor.BackColor=this.colorDialog.Color;
            //设置符号颜色为用户选定的颜色
            switch (this.axSymbologyControl.StyleClass)
            {
                //点符号
                 case esriSymbologyStyleClass.esriStyleClassMarkerSym-
bols:
                    IMarkerSymbol mSymbol=(IMarkerSymbol)this.m_pStyle-
GalleryItem.Item;
                    mSymbol.Color=this.ColorToIRgbColor(this.colorDia-
log.Color);
                    break;
                //线符号
                case esriSymbologyStyleClass.esriStyleClassLineSymbols:
                    ILineSymbol lSymbol=(ILineSymbol)this.m_pStyleGal-
leryItem.Item;
                    lSymbol.Color=this.ColorToIRgbColor(this.colorDia-
log.Color);
                    break;
                //面符号
                case esriSymbologyStyleClass.esriStyleClassFillSymbols:
                    IFillSymbol fSymbol=(IFillSymbol)this.m_pStyleGal-
leryItem.Item;
                    fSymbol.Color=this.ColorToIRgbColor(this.colorDia-
log.Color);
                    break;
            }
            //更新符号预览
            this.PreviewImage();
        }
    }

    private void btnOutlineColor_Click(object sender,EventArgs e)
```

```
    {
        //调用系统颜色对话框
        if (this.colorDialog.ShowDialog()= = DialogResult.OK)
        {
            //将颜色按钮的背景颜色设置为用户选定的颜色
            this.btnColor.BackColor=this.colorDialog.Color;
            //设置符号颜色为用户选定的颜色
            switch (this.axSymbologyControl.StyleClass)
            {
                //面符号
                case esriSymbologyStyleClass.esriStyleClassFillSymbols:
                    IFillSymbol pSymbol=this.m_pStyleGalleryItem.Item
as IFillSymbol;
                    pSymbol.Outline.Color=this.ColorToIRgbColor(this.
colorDialog.Color);
                    break;
            }
            //更新符号预览
            this.PreviewImage();
        }
    }
```

至此，已经能够修改符号的参数。

3.5.2 添加更多符号菜单

单击“更多符号”按钮，弹出的菜单(contextMenuStripMoreSymbol)中列出了 ArcGIS 自带的其他符号，勾选相应的菜单项就可以在 SymbologyControl 中增加相应的符号。菜单的最后一项是“添加符号”，选择这一项时，将弹出打开文件对话框，我们可以由此选择其他的 ServerStyle 文件，以加载更多的符号。

(1)定义全局变量

在 SymbolSelectorFrm 中定义如下全局变量，用于判断菜单是否已经初始化。

```
//菜单是否已经初始化标志
bool contextMenuMoreSymbolInitiated=false;
```

(2)双击“更多符号”按钮，添加 Click 事件。

在此事件响应函数中，我们要完成 ServerStyle 文件的读取，将其文件名作为菜单项名称生成菜单并显示菜单。代码如下：

```
private void btnMoreSymbols_Click(object sender,EventArgs e)
{
    if (this.contextMenuMoreSymbolInitiated= = false)
```

```
        {
            string sInstall = ESRI.ArcGIS.RuntimeManager.ActiveRuntime.
Path;
            string path=System.IO.Path.Combine(sInstall,"Styles");
            //取得菜单项数量
            string[] styleNames=System.IO.Directory.GetFiles(path,"*.
ServerStyle");
            ToolStripMenuItem[] symbolContextMenuItem=
                                    new ToolStripMenuItem[st-
yleNames.Length+1];
            //循环添加其他符号菜单项到菜单
            for (int i=0; i < styleNames.Length; i++)
            {
                symbolContextMenuItem[i]=new ToolStripMenuItem();
                symbolContextMenuItem[i].CheckOnClick=true;
                symbolContextMenuItem[i].Text=
                            System.IO.Path.GetFileNameWithoutEx-
tension(styleNames[i]);
                if (symbolContextMenuItem[i].Text== "ESRI")
                {
                    symbolContextMenuItem[i].Checked=true;
                }
                symbolContextMenuItem[i].Name=styleNames[i];
            }
            //添加"更多符号"菜单项到菜单最后一项
            symbolContextMenuItem[styleNames.Length] = new ToolStrip-
MenuItem();
            symbolContextMenuItem[styleNames.Length].Text="添加符号";
            symbolContextMenuItem[styleNames.Length].Name="AddMoreSymbol";
            //添加所有的菜单项到菜单
            this.contextMenuStripMoreSymbol.Items.AddRange(symbolCon-
textMenuItem);
            this.contextMenuMoreSymbolInitiated=true;
        }
        //显示菜单
        this.contextMenuStripMoreSymbol.Show(this.btnMoreSymbols.Lo-
cation);
    }
```

(3)添加 contextMenuStripMoreSymbol 控件的 ItemClicked 事件

当单击某一菜单项时响应 ItemClicked 事件，将选中的 ServerStyle 文件导入到 SymbologyControl 中并刷新。当用户单击“添加符号”菜单项时，弹出打开文件对话框，供用户选择其他的 ServerStyle 文件。代码如下：

```
/// “更多符号”按钮弹出的菜单项单击事件
    private void contextMenuStripMoreSymbol_ItemClicked(object sender,
                                                                ToolStripItemCli-
ckedEventArgs e)
    {
         ToolStripMenuItem pToolStripMenuItem = (ToolStripMenuItem) e.
ClickedItem;
        //如果单击的是“添加符号”
        if (pToolStripMenuItem.Name = = "AddMoreSymbol")
        {
            //弹出打开文件对话框
            if (this.openFileDialog.ShowDialog() = = DialogResult.OK)
            {
                //导入 style file 到 SymbologyControl
                this.axSymbologyControl.LoadStyleFile(this.openFileDi-
alog.FileName);
                //刷新 axSymbologyControl 控件
                this.axSymbologyControl.Refresh();
            }
        }
        else//如果是其他选项
        {
            if (pToolStripMenuItem.Checked = = false)
            {
                 this.axSymbologyControl.LoadStyleFile(pToolStripMenu-
Item.Name);
                this.axSymbologyControl.Refresh();
            }
            else
            {
                this.axSymbologyControl.RemoveFile(pToolStripMenuItem.
Name);
                this.axSymbologyControl.Refresh();
            }
```

```
        }
    }
```

3.6 编译运行

按下 F5 键，编译程序运行。

第4章　图层标注

4.1　知识要点

在 ArcGIS Engine 中，可以用更复杂的方法对要素图层进行标注。涉及 ILabelEngine-LayerProperties、IAnnotateLayerPropertiesCollection、IAnnotateLayerProperties 等接口。

LabelEngineLayerProperties 是与某个要素图层关联的，用于描述要素图层的标注 Labe-lEngineLayerProperties 类实现了 ILabelEngineLayerProperties 接口，主要属性如下：

①Expression 属性用于通过 VBScript 或 JAScript 设置标注表达式或格式化标注字段；

②BasicOverposterLayerProperties 属性用于设置标注位置，并有处理标注冲突的功能；

③Symbol 属性用于设置标注字体的格式。

AnnotateLayerPropertiesCollection 是一个要素图层的属性，可自 IGeoFeaturelayer 的 AnnotationProperties属性获取，它是标注对象 LabelEngineLayerProperties 的集合。

4.2　功能描述

在图层操作浮动菜单上点击“Layer Label”菜单项，弹出“图层标注”对话框，如图 4-1 所示。

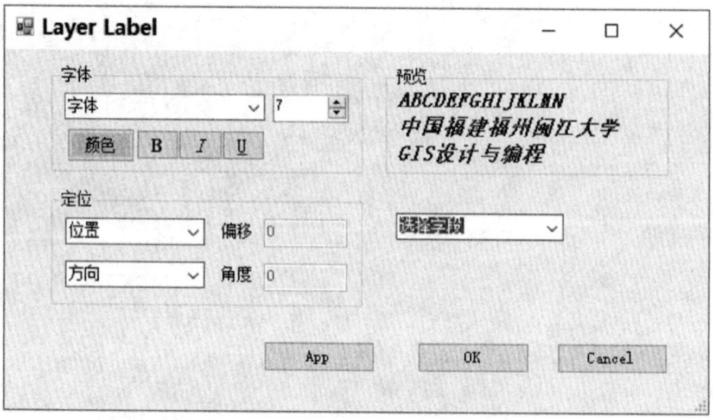

图 4-1　“图层标准”对话框

4.3　功能实现

1. 功能类设计

新建一个 Windows 窗体，命名为"LabelLayerFrm. cs"。从工具箱拖动表 4-1 列出的控件到窗体。

表 4-1　　控件及其属性、含义

控　　件	Name 属性	含　义	其　他
Combox	cbxField	标注字段	
Combox	cbxFont	字体	
NumericUpdown	nudSize	字体大小	
Button	btnColor	字体颜色	
Button	btnBold	粗体	
Button	btnItalic	斜体	
Button	btnUnderline	下画线	
Combox	cbxPosition	标注位置	
Combox	cbxOrentation	标注方向	
TextBox	txtOffset	标注偏移	
TextBox	txtAngle	标准角度	
RichTextBox	rtxtPreview	预览	
Button	btnApp	应用	
Button	btnOK	确定	DialogRezult. OK
Button	btnCancel	取消	

添加如下代码：

```
public partial class LabelLayerFrm:Form
{
    public IFeatureLayer m_pLayer;
    public LabelLayerFrm(ILayer layer)
    {
        InitializeComponent();
        m_pLayer=layer as IFeatureLayer;
    }
    //事件响应函数
```

```
        private void LabelLayerFrm_Load(object sender,EventArgs e)
        private void btnColor_Click(object sender,EventArgs e)
        private void btnBold_Click(object sender,EventArgs e)
        private void btnItalic_Click(object sender,EventArgs e)
        private void btnUnderline_Click(object sender,EventArgs e)
        private void nudSize_ValueChanged(object sender,EventArgs e)
        private void btnOK_Click(object sender,EventArgs e)
        private void btnApp_Click(object sender,EventArgs e)
        //核心功能函数
        private void LayerLabel()
        //辅助函数
        private esriOverposterPointPlacementMethod getPointPlacement-
Method()
        private ILineLabelPosition getLineLabelPosition()
        private esriOverposterPolygonPlacementMethod getPolygonOrenta-
tionMethod()
        private System.Drawing.Font CreateFont()
        private void Preview()
        public IColor ColorToIRgbColor(Color color)
    }
```

2. 实现响应函数

代码如下：

```
//窗体加载时,填充字段下拉框,填充字体下拉框,填充标注位置和方向下拉框
private void LabelLayerFrm_Load(object sender,EventArgs e)
{
    //加载图层字段
    ITable pTable=m_pLayer as ITable;
    IField pField=null;
    for (int i=0; i < pTable.Fields.FieldCount; i++)
    {
        pField=pTable.Fields.get_Field(i);
        cbxField.Items.Add(pField.AliasName);
    }

    //填充安装字体
    InstalledFontCollection fc=new InstalledFontCollection();
    foreach (FontFamily font in fc.Families)
    {
```

```
        this.cbxFont.Items.Add(font.Name);
    }

    //设置标注位置和方向
    switch(m_pLayer.FeatureClass.ShapeType)
    {
        case esriGeometryType.esriGeometryPoint:
            this.cbxPosition.Items.Add("AroundPoint");
            this.cbxPosition.Items.Add("OnTopPoint");
            this.cbxPosition.Items.Add("SpecifiedAngles");
            this.cbxPosition.Items.Add("RotationField");
            break;
        case esriGeometryType.esriGeometryPolyline:
            this.cbxPosition.Items.Add("At Start");
            this.cbxPosition.Items.Add("At End");
            this.cbxPosition.Items.Add("InLine");

            this.cbxOrentation.Items.Add("Horizontal");
            this.cbxOrentation.Items.Add("Perpendicular");
            this.cbxOrentation.Items.Add("Parallel");
            this.cbxOrentation.Items.Add("Curved");

            this.txtAngle.Enabled=false;
            break;
        case esriGeometryType.esriGeometryPolygon:
            this.cbxPosition.Items.Add("OnlyInsidePolygon");
            this.cbxPosition.Items.Add("  ");

            this.cbxOrentation.Items.Add("AlwaysHorizontal");
            this.cbxOrentation.Items.Add("AlwaysStraight");
            this.cbxOrentation.Items.Add("MixedStrategy");

            this.txtOffset.Enabled=false;
            this.txtAngle.Enabled=false;
            break;
    }

    Preview();
```

```
    }

    //字体、颜色、大小选择现响应函数
    private void cbxFont_SelectedIndexChanged(object sender,EventArgs
e)
    {
        Preview();
    }

    private void btnColor_Click(object sender,EventArgs e)
    {
        if (this.colorDialog1.ShowDialog()== DialogResult.OK)
        {
            this.btnColor.BackColor=colorDialog1.Color;
            Preview();
        }
    }

    private void btnBold_Click(object sender,EventArgs e)
    {
        if (btnBold.BackColor== Color.AntiqueWhite)
            btnBold.BackColor=btnOK.BackColor;
        else if (btnBold.BackColor== btnOK.BackColor)
            btnBold.BackColor=Color.AntiqueWhite;

        Preview();
    }

    private void btnItalic_Click(object sender,EventArgs e)
    {
        if (btnItalic.BackColor== Color.AntiqueWhite)
            btnItalic.BackColor=btnOK.BackColor;
        else if (btnItalic.BackColor== btnOK.BackColor)
            btnItalic.BackColor=Color.AntiqueWhite;

        Preview();
    }
```

```
private void btnUnderline_Click(object sender,EventArgs e)
{
    if (btnUnderline.BackColor == Color.AntiqueWhite)
        btnUnderline.BackColor=btnOK.BackColor;
    else if (btnUnderline.BackColor == btnOK.BackColor)
        btnUnderline.BackColor=Color.AntiqueWhite;

    Preview();
}

private void nudSize_ValueChanged(object sender,EventArgs e)
{
    Preview();
}
//确定,应用响应函数
private void btnOK_Click(object sender,EventArgs e)
{
    if (btnApp.Enabled == true)
    {
        btnApp_Click(null,null);
    }
    this.Close();
}

private void btnApp_Click(object sender,EventArgs e)
{
    if (this.cbxField.SelectedIndex < 0 || cbxFont.SelectedIndex <
0)
    {
        MessageBox.Show("请选择标注字段或字体");
        return;
    }
    LayerLabel();
}
```

注意：控制字体的黑体、斜体、下画线三种属性的按钮，背景颜色在两个颜色中切换：它们是 AntiqueWhite，btnOK 的背景色。呈现 AntiqueWhite 颜色表示该属性被选中。

3. 核心函数实现

LayerLabel(...)是实现标注的总调度函数，步骤如下：

①清空默认注记属性;

②创建标注引擎：包括配置表达式，配置位置属性，配置文本符号;

③标注引擎添加到注记属性集。

代码如下:

```
private void LayerLabel()
{
    //清空默认注记属性
    IGeoFeatureLayer pGeoFeatureLayer=m_pLayer as IGeoFeatureLayer;
    pGeoFeatureLayer.AnnotationProperties.Clear();

    //创建标注引擎
    ILabelEngineLayerProperties pLableEngine = new LabelEngineLay-
erPropertiesClass();
    {
        //配置表达式
        string pLable="["+cbxField.SelectedItem.ToString()+"]";
        pLableEngine.Expression=pLable;
        pLableEngine.IsExpressionSimple=true;

        //配置位置属性
        IBasicOverposterLayerProperties4 pBasic4 = new BasicOver-
posterLayerPropertiesClass();
        pBasic4.NumLabelsOption=esriBasicNumLabelsOption.esriO-
neLabelPerShape;
        switch (m_pLayer.FeatureClass.ShapeType)
        {
            case esriGeometryType.esriGeometryPoint:
                pBasic4.PointPlacementMethod=this.getPointPlace-
mentMethod();
                pBasic4.BufferRatio=double.Parse(txtOffset.Text);
                pBasic4.PointPlacementAngles=new double[1] { doub-
le.Parse(txtAngle.Text) };
                break;
            case esriGeometryType.esriGeometryPolyline:
                pBasic4.LineLabelPosition=this.getLineLabelPosi-
tion();
                pBasic4.LineOffset=double.Parse(txtOffset.Text);
                break;
```

```
                case esriGeometryType.esriGeometryPolygon:
                    pBasic4.PlaceOnlyInsidePolygon=(this.cbxPosition.
SelectedIndex==0)?true:
                    false;
                    pBasic4.PolygonPlacementMethod=this.getPolygonOren-
tationMethod();
                    break;
            }
            pLableEngine.BasicOverposterLayerProperties = pBasic4 as
IBasicOverposterLayerProperties;

            //配置文本符号
            ITextSymbol pTextSymbol=new TextSymbolClass();
            pTextSymbol.Font=OLE.GetIFontDispFromFont(CreateFont()) as
IFontDisp;
            pTextSymbol.Color = ColorToIRgbColor(this.btnColor.Back-
Color);
            pLableEngine.Symbol=pTextSymbol;
        }

        //标注引擎属性添加到主机属性集
        pGeoFeatureLayer.AnnotationProperties.Add(pLableEngine as
IAnnotateLayerProperties);
        pGeoFeatureLayer.DisplayAnnotation=true;
    }
```

4. 辅助函数

代码如下：

```
    //点类型定位方法
    private esriOverposterPointPlacementMethod getPointPlacementMethod()
    {
        switch (this.cbxPosition.SelectedItem.ToString())
        {
            case "AroundPoint":
            default:
                return esriOverposterPointPlacementMethod.esriAroundP-
oint;
            case "OnTopPoint":
                return esriOverposterPointPlacementMethod.esriOnTop-
```

```
Point;
            case "RotationField":
                return esriOverposterPointPlacementMethod.esriRotation-
Field;
            case "SpecifiedAngles":
                return esriOverposterPointPlacementMethod.esriSpeci-
fiedAngles;
        }
    }
    //线类型定位和角度
    private ILineLabelPosition getLineLabelPosition()
    {
        ILineLabelPosition pLinePos=new LineLabelPositionClass();
        switch(this.cbxPosition.SelectedItem.ToString())
        {
            case "At Start":
                pLinePos.AtStart=true;
                pLinePos.AtEnd=false;
                pLinePos.InLine=false;
                break;
            case "At End":
                pLinePos.AtStart=false;
                pLinePos.AtEnd=true;
                pLinePos.InLine=false;
                break;
            case "InLine":
            default:
                pLinePos.AtStart=false;
                pLinePos.AtEnd=false;
                pLinePos.InLine=true;
                break;
        }
        switch (this.cbxOrentation.SelectedItem.ToString())
        {
            case "Horizontal":
                pLinePos.Horizontal=true;
                pLinePos.Perpendicular=false;
                pLinePos.Parallel=false;
```

```
                pLinePos.ProduceCurvedLabels=false;
                break;
            case "Perpendicular":
                pLinePos.Horizontal=false;
                pLinePos.Perpendicular=true;
                pLinePos.Parallel=false;
                pLinePos.ProduceCurvedLabels=false;
                break;
            case "Parallel":
                pLinePos.Horizontal=false;
                pLinePos.Perpendicular=false;
                pLinePos.Parallel=true;
                pLinePos.ProduceCurvedLabels=false;
                break;
            case "Curved":
            default:
                pLinePos.Horizontal=false;
                pLinePos.Perpendicular=false;
                pLinePos.Parallel=true;      //必须为真
                pLinePos.ProduceCurvedLabels=true;
                break;
        }

        return pLinePos;
    }
    //面类型定位方向
    private  esriOverposterPolygonPlacementMethod  getPolygonOrenta-
tionMethod()
    {
        switch (this.cbxOrentation.SelectedItem.ToString())
        {
            case "AlwaysStraight":
                return esriOverposterPolygonPlacementMethod.esriAlway-
sStraight;
            case "MixedStrategy":
                return esriOverposterPolygonPlacementMethod.esriMixed-
Strategy;
            case "AlwaysHorizontal":
```

```
            default:
                return esriOverposterPolygonPlacementMethod.esriAlway-
sHorizontal;
        }
    }
    //创建字体
    private System.Drawing.Font CreateFont( )
    {
        string fontFamilyName=this.cbxFont.SelectedItem.ToString( );
        int size=(int)this.nudSize.Value;

        FontStyle fontStyle=FontStyle.Regular;
        if(btnItalic.BackColor== Color.AntiqueWhite)
            fontStyle=fontStyle | FontStyle.Italic;

        if(btnBold.BackColor== Color.AntiqueWhite)
            fontStyle=fontStyle | FontStyle.Bold;

        if(btnUnderline.BackColor== Color.AntiqueWhite)
            fontStyle=fontStyle | FontStyle.Underline;

        System.Drawing.Font font = new System.Drawing.Font(fontFami-
lyName,size,fontStyle);
        return font;
    }
    //颜色转换
    public IColor ColorToIRgbColor(Color color)
    {
        IColor pColor=new RgbColorClass();
        pColor.RGB=color.B*65536+color.G*256+color.R;
        return pColor;
    }
    //预览
    private void Preview()
    {
        if (this.cbxFont.SelectedIndex >= 0)
        {
            this.rtxtPreview.ForeColor=this.btnColor.BackColor;
```

```
            this.rtxtPreview.Font =CreateFont();
        }
    }
```

4.4 调用图层标注窗体

图层浮动菜单上添加 Label Layer 菜单项，建立如下响应函数：

```
    private void labelLayerToolStripMenuItem_Click(object sender,Even-
tArgs e)
    {
        LabelLayerFrm labelLyrFrm=new LabelLayerFrm(m_tocRightLayer);
        if (labelLyrFrm.ShowDialog()= = DialogResult.OK)
        {
            _mapControl.Refresh(esriViewDrawPhase.esriViewGraphics,
null,null);
        }
    }
```

4.5 编译运行

按下 F5 键，编译运行程序。

第5章　唯一值渲染

5.1　知识要点

专题图可以突出地图要表达的信息，ArcGIS Engine 的 FeatureRenderer 对象为用户提供了多个着色方案用于制作专题图，在这个过程中既可以使用标准的着色方案，也可以定制自己的着色方案。

FeatureRenderer 是一个抽象类，其子类负责进行不同类型的着色运算，包括：简单绘制(SimpleRenderer)；唯一值绘制(Unique ValueRenderer)或多字段唯一值绘制；点密度或多字段点密度绘制(DotDensityRenderer)；数据分级绘制(ClassBreaksRenderer)；饼图或直方图(ChartRenderer)；比例符号渲染(ScaleDependentRenderer)。它们都实现了 IFeatureRenderer 接口，这个接口定义了进行地图着色运算的公共属性和方法。

而渲染对象是要素图层的一个属性，程序员可以通过 IGeoFeatureLayer. Renderer 属性(是一个可读写属性)，获得一个图层的渲染对象或为其赋值。

唯一值绘制方法是依据要素图层的要素类中的字段值，对每个字段值分别进行渲染，使用唯一值绘制可以通过颜色区分每个要素。UniqueValueRenderer 对象类实现了 IUniqueValueRenderer 接口，该接口定义了几个重要属性和方法。

①Field：提供唯一分类值的字段。

②Value：特征的唯一分类值。

③ValueCount：需要显示的唯一分类值的数目 。

④AddValue (string Value, string Heading, ISymbol Symbol)：添加"值-符号"对。

UniqueValueRenderer 要素类渲染的使用方法如下：

①先创建一个 UniqueValueRenderer 对象；

②然后遍历图层中的所有要素，获取指定字段的唯一值集合；

③使用 UniqueValueRenderer 的 AddValue( )为每个唯一值匹配不同颜色或符号；

④然后将其绑定到图层。

5.2　功能描述

点击【Theme】Tab 页上【Unique Value】按钮，弹出"唯一值渲染"对话框，如图 5-1 所示，可选择图层、渲染字段、配色方案等。

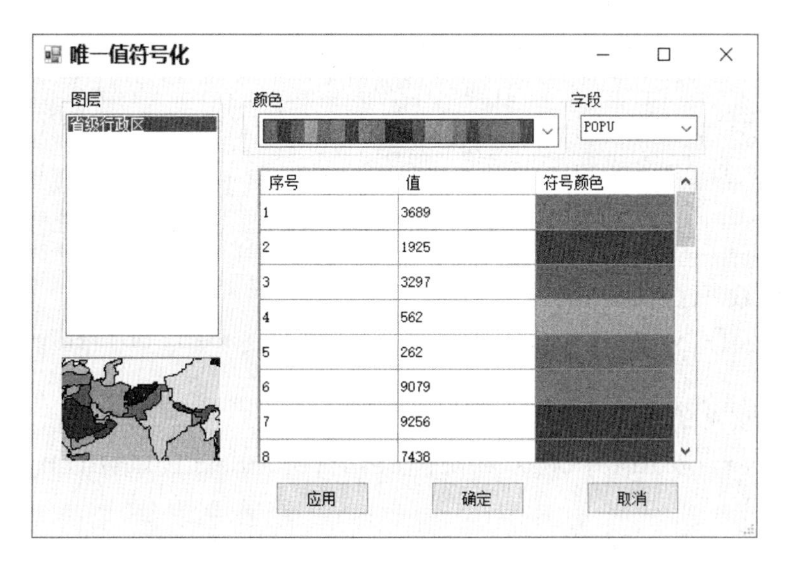

图 5-1　“唯一值符号化”对话框

5.3　功能实现

1. 功能类设计

新建一个 Windows 窗体，命名为“UniqueValueRendererFrm. cs”。

从工具箱拖一个 ListBox，两个 ComBox(图层列表，选择字段)，其中一个改为自定义派生类 ComBoxEx、一个 DataGridView，三个 Button 控件到窗体。

控　　件	Name 属性	Text 属性	其　他
ListBox	ListBoxLayers	选择图层	
Combox	cbxSelField	选择字段	
ComboxEx	imgcbxColorRamp	色带选择	自定义派生类
DataGridView	dataGridView1	显示值符号对	
Button	btnApp	应用	
Button	btnOK	确定	DialogRezult. OK
Button	btnCancel	取消	

添加如下引用代码：

```
public partial class UniqueValueRendererFrm:Form
{
```

```
        IMapControl3 m_mapControl=null;
        public UniqueValueRendererFrm(IMapControl3 mapControl)
        {
            InitializeComponent();
            m_mapControl=mapControl;
        }

        //装载事件响应函数:
        private void UniqueValueRendererFrm_Load(object sender,Even-
tArgs e)
        //图层选择响应函数:
        private void ListBoxLayers_SelectedIndexChanged(object sender,
EventArgs e)
        //字段选择响应函数:
        private void cbxSelFields_SelectedIndexChanged(object sender,
EventArgs e)
        //色带选择响应函数:
        private void imgcbxColorRamp_SelectedIndexChanged(object send-
er,EventArgs e)
        //确定按钮响应函数:
        private void btnOK_Click(object sender,EventArgs e)
        //应用按钮响应函数:
        private void btnApply_Click(object sender,EventArgs e)

        //核心函数:获取唯一值集合
        public IEnumerator GetUniqueValues(IFeatureLayer pfeaturelayer,
string strFldName,out int nUValueCount)
        //核心函数:唯一值渲染控制
        private void UniqueValueRenderer(IFeatureLayer pFeatLyr,string
sFieldName)

        //若干辅助函数
        ……
    }
```

2. 实现响应函数

代码如下:

```
    //窗体加载时,填充图层列表框,初始化色带 ConboxEx
    private void UniqueValueRenderFrm_Load(object sender,EventArgs e)
```

```
    {
        IEnumLayer layers=GetLayers();
        layers.Reset();
        ILayer layer=null;
        while ((layer=layers.Next())!= null)
        {
            ListBoxLayers.Items.Add(layer.Name);
        }

        //设置色带
        //Get the ArcGIS install location
        string sStyleFile=ESRI.ArcGIS.RuntimeManager.ActiveRuntime.Path;
        sStyleFile += "Styles\\ESRI.ServerStyle";
        InitColorRamp(sStyleFile);

        if (ListBoxLayers.Items.Count > 0)
        {
            ListBoxLayers.SelectedIndex=0;
            imgcbxColorRamp.SelectedIndex=0;
        }
    }

    //图层选定后,响应函数将该层适合唯一值符号化的字段名填充到字段下拉框
    private void ListBoxLayers_SelectedIndexChanged(object sender,
EventArgs e)
    {
        cbxSelFields.Items.Clear();
        cbxSelFields.Text="";
        IField pField=null;

         IFeatureLayer pFtLayer = GetFeatureLayer(ListBoxLayers.Selecte-
dItem.ToString());
        IFeatureClass pFeatCls=pFtLayer.FeatureClass;
        for (int i=0; i < pFeatCls.Fields.FieldCount; i++)
        {
            pField=pFeatCls.Fields.get_Field(i);
            if (pField.Type== esriFieldType.esriFieldTypeDouble ||
                pField.Type== esriFieldType.esriFieldTypeInteger ||
```

```
                pField.Type == esriFieldType.esriFieldTypeSingle ||
                pField.Type == esriFieldType.esriFieldTypeSmallInteger)
            {
                if (!cbxSelFields.Items.Contains(pField.Name))
                {
                    cbxSelFields.Items.Add(pField.Name);
                }
            }
        }
    }
    //色带选择响应函数
    //选定色带后,在 DataGridView 颜色列中,将单元背景色设置为指定色带的相应值
    private void imgcbxColorRamp_SelectedIndexChanged(object sender,
EventArgs e)
    {
        if(imgcbxColorRamp.SelectedIndex < 0 )
                return;

        //将色带颜色值显示在 DataGridView 中
        IEnumColors pColorEnumerater = ((ItemEx)imgcbxColorRamp.Select-
edItem).ColorEnumerater;
        pColorEnumerater.Reset();

        for (int i=0; i < this.dataGridView1.RowCount; i++)
        {
            IColor pNextUniqueColor = pColorEnumerater.Next();
            if (pNextUniqueColor == null)
            {
                pColorEnumerater.Reset();
                pNextUniqueColor = pColorEnumerater.Next();
            }

            this.dataGridView1[2,i].Style.BackColor = ColorTranslator.
FromOle(pNextUniqueColor.RGB);
        }
    }
    //字段选择响应函数:
```

①获取字段唯一值集合(用到 GetUniqueValues()函数);

②显示字段值(用到 DisplayValues);

③激活色带选择响应函数;

```
    private void cbxSelFields_SelectedIndexChanged(object sender,Even-
tArgs e)
    {
        //获取唯一值集合枚举器
        string fieldName=this.cbxSelFields.SelectedItem.ToString();
        string layerName=this.ListBoxLayers.SelectedItem.ToString();
        IFeatureLayer pFeatureLayer=this.GetFeatureLayer(layerName);
        int nUValueCount;
        IEnumerator pEnumVariantSimple = GetUniqueValues(pFeatureLayer,
fieldName,out nUValueCount);
        if (nUValueCount > 4000)
        {
            MessageBox.Show("当前需要生成的符号总数太多(大于 3000),无法显
示");
            btnApply.Enabled=false;
        }
        else
        {
            //重现填充 DataGridView 值列
            DisplayValues(pEnumVariantSimple);

            //激活色带选择响应函数;
            imgcbxColorRamp_SelectedIndexChanged(sender,e);
            btnApply.Enabled=true;
        }
    }

    //对选定图层和字段调用 UniqueValueRenderer(...)函数:
    private void btnApply_Click(object sender,EventArgs e)
    {
        IFeatureLayer pFeatLyr = GetFeatureLayer(ListBoxLayers.Selecte-
dItem.ToString());
        string sFieldName=cbxSelFields.SelectedItem.ToString();
        UniqueValueRenderer(pFeatLyr,sFieldName);
    }
```

```
    private void btnOk_Click(object sender,EventArgs e)
    {
        if (btnApply.Enabled== true)
        {
            btnApply_Click(null,null);
        }
        this.Close();
    }
```

3. 核心函数实现

GetUniqueValues(...)获取图层中渲染字段的唯一值的集合，用 IDataStatistics 接口实现，代码如下：

```
    public IEnumerator GetUniqueValues ( IFeatureLayer pfeaturelayer,
string strFldName,out int nUValueCount)
    {
        IEnumerator functionReturnValue=null;
        nUValueCount=0;
        try
        {
            //创建过滤器
            IQueryFilter pQueryFilter=new QueryFilter();
            pQueryFilter.SubFields=strFldName;

            //创建游标(结果只有一个字段)
            IFeatureClass pFeatureClass=pfeaturelayer.FeatureClass;
            IFeatureCursor pFeatureCursor=pFeatureClass.Search(pQuery-
Filter,true);

            //创建数据统计对象
            IDataStatistics pDastStat=new DataStatistics();
            {
                pDastStat.Field=strFldName;
                pDastStat.Cursor=(ICursor)pFeatureCursor;
            }

            //取得唯一值集合
            functionReturnValue=pDastStat.UniqueValues;
            nUValueCount=pDastStat.UniqueValueCount;
```

```
            functionReturnValue.Reset();
            return functionReturnValue;
        }
        catch (Exception ex)
        {
            MessageBox.Show(ex.ToString());
            return null;
        }
    }
```

UniqueValueRenderer()是唯一值渲染的调度函数，步骤如下：

①创建唯一值渲染器；

②根据 DataGridView 表，为每个“值”配置一个符号；

③为要素类设置渲染器。

这里用到 CreateDefinedSymbol()：根据要素的集合类型创建相应的类型符号。代码如下：

```
    private void UniqueValueRenderer ( IFeatureLayer pFeatLyr, string
sFieldName)
    {
        //创建唯一值渲染器：
         IUniqueValueRenderer pUniqueValueRender = new UniqueValueRen-
dererClass();
        pUniqueValueRender.FieldCount =1;//设置唯一值符号化的关键字段为
一个
        pUniqueValueRender.set_Field(0,sFieldName);//设置唯一值符号化的
第一个关键字段
        try
        {
            //为每个字段值配置颜色
            esriGeometryType shpType=pFeatLyr.FeatureClass.ShapeType;
            for (int i=0; i < dataGridView1.RowCount; i++)
            {
                object vntUniqueValue=dataGridView1[1,i].Value;
                Color uniqueColor=this.dataGridView1[2,i].Style.Back-
Color;

                ISymbol pSymbol=CreateDefinedSymbol(shpType,ColorToIR-
gbColor(uniqueColor));
```

```
                pUniqueValueRender.AddValue(vntUniqueValue.ToString
(),"",pSymbol);
            }
        }
        catch (Exception ex)
        {
        }
        //将渲染器赋值给要素层
        IGeoFeatureLayer pGeoFeatLyr=pFeatLyr as IGeoFeatureLayer;
        pGeoFeatLyr.Renderer=pUniqueValueRender as IFeatureRenderer;
    }
```

4. 辅助函数

```
    //将某一字段的唯一值显示在 DataGridView
    private void DisplayValues(IEnumerator pEnumVariantSimple)
    {
        //将属性值填充到表格的第一列
        dataGridView1.Rows.Clear();
        pEnumVariantSimple.Reset();

        object vntUniqueValue=null;
        int i=0;
        while( pEnumVariantSimple.MoveNext() )
        {
            vntUniqueValue=pEnumVariantSimple.Current;
            {
                dataGridView1.Rows.Add();
                dataGridView1[0,i].Value=(i+1).ToString();
                dataGridView1[1,i].Value=vntUniqueValue;
            }
            i++;
        }

        dataGridView1.AllowUserToAddRows=false;
    }
    //初始化色带 ComBoxEx 控件(使用 ArcEngine 自带符号库中的随机色带)
    private void InitColorRamp(string sStyleFile)
    {
        IStyleGallery styleGallery=new ServerStyleGalleryClass();
```

```
        IStyleGalleryStorage styleGalleryStorage = styleGallery as IStyle-
GalleryStorage;
        styleGalleryStorage.AddFile(sStyleFile);

        //styleGallery.LoadStyle(sInstall,"Color Ramps");
        IEnumStyleGalleryItem enumStyleGalleryItem = styleGallery.get_
Items("Color Ramps",
       sStyleFile,"");
        enumStyleGalleryItem.Reset();

        //填充 ComboBoxEx
        IStyleGalleryItem styleGalleryItem=enumStyleGalleryItem.Next();
        while (styleGalleryItem!= null)
        {
            IColorRamp pColorRamp=styleGalleryItem.Item as IColorRamp;
            if (pColorRamp is IRandomColorRamp)
            {
                pColorRamp.Size=100;
                bool createRamp;
                pColorRamp.CreateRamp(out createRamp);
                imgcbxColorRamp.Items.Add(new ItemEx(pColorRamp.Colors));
            }

            styleGalleryItem=enumStyleGalleryItem.Next();
        }

        //Remove Files
        styleGalleryStorage.RemoveFile(sStyleFile);
        System.Runtime.InteropServices.Marshal.ReleaseComObject(enum-
StyleGalleryItem);
        System.Runtime.InteropServices.Marshal.ReleaseComObject(style-
Gallery);

        imgcbxColorRamp.isUniqueValue=true;
    }
    //创建特定符号
    private ISymbol CreateDefinedSymbol(esriGeometryType shpType,
IColor pNextUniqueColor)
```

```
{
    ISymbol pISymbol=null;
    switch (shpType)
    {
        case esriGeometryType.esriGeometryPolygon:
        {
            IFillSymbol pFillSymbol=new SimpleFillSymbolClass();
            pFillSymbol.Color=pNextUniqueColor;
            pISymbol=pFillSymbol as ISymbol;
            break;
        }
        case esriGeometryType.esriGeometryPolyline:
        {
            ILineSymbol pLineSymbol=new SimpleLineSymbolClass();
            pLineSymbol.Color=pNextUniqueColor;
            pISymbol=pLineSymbol as ISymbol;
            break;
        }
        case esriGeometryType.esriGeometryPoint:
        {
            IMarkerSymbol pMarkerSymbol=new SimpleMarkerSymbolClass();
            pMarkerSymbol.Color=pNextUniqueColor;
            pISymbol=pMarkerSymbol as ISymbol;
            break;
        }
    }
    return pISymbol;
}
```

此外，还涉及两个 GetFeatureLayer( )和 GetLayers( )，ColorToIRgbColor( )三个辅助函数，请参阅之前有关内容。

5.4 ComboBoxEx 派生类

为实现 ComboBox 控件支持色带显示，编写 ComboBox 派生类 ComboBoxEx，重点是重载 OnDrawItem(...)函数，这里支持渐变色带和随机色带，有 isUniqueValue 属性控制。

代码如下：

```
public class ComboBoxEx:ComboBox
{
```

```
        public bool isUniqueValue { get; set; }
        public ComboBoxEx()
        {
            DrawMode=DrawMode.OwnerDrawFixed;
            DropDownStyle=ComboBoxStyle.DropDownList;
            ItemHeight=20;
            Width=240;
        }

        protected override void OnDrawItem(DrawItemEventArgs e)
        {
            if (e.Index < 0)
                return;

            //获取范围矩形
            Rectangle rect=e.Bounds;
            if (!isUniqueValue)
            {
                //读取起始、终止颜色值
                IColor fromColor=((ItemEx)Items[e.Index]).FromColor;
                IColor toColor=((ItemEx)Items[e.Index]).ToColor;
                Color _FromColor=ColorTranslator.FromOle(fromColor.RGB);
                Color _ToColor=ColorTranslator.FromOle(toColor.RGB);

                //选择线性渐变刷子
                 LinearGradientBrush brush=new LinearGradientBrush(rect,_
FromColor,_ToColor,0,false);
                //矩形缩小一个像素;
                rect.Inflate(-1,-1);

                //填充颜色
                e.Graphics.FillRectangle(brush,rect);
                //绘制边框
                e.Graphics.DrawRectangle(Pens.Black,rect);
            }
            else
            {
                //获取颜色枚举器
```

```
                IEnumColors pColorEnumerater = ((ItemEx) Items[e.Index]).
ColorEnumerater;
                pColorEnumerater.Reset();
                //矩形缩小一个像素;
                rect.Inflate(-1,-1);

                //按 1/10 宽度绘制颜色
                float f=rect.Width/20.0f;
                for(int i=0; i < 20; i++)
                {
                    //新建单一色刷子,颜色为对应项记录的值
                    IColor pNextColor=pColorEnumerater.Next();
                    Color pColor=ColorTranslator.FromOle(pNextColor.RGB);
                    SolidBrush brush=new SolidBrush(pColor);

                    e.Graphics.FillRectangle(brush,rect.X+f*i,rect.Y,
f,rect.Height);
                }

                //绘制边框
                e.Graphics.DrawRectangle(Pens.Black,rect);
            }
        }

        public class ItemEx
        {
            public ItemEx(IColor fromClaor,IColor toColor)
            {
                FromColor=fromClaor;
                ToColor=toColor;
            }
            public ItemEx(IEnumColors pColors)
            {
                ColorEnumerater=pColors;
            }
            public IColor FromColor { get; set; }
            public IColor ToColor { get; set; }
            public IEnumColors ColorEnumerater { get; set; }
```

```
        }
    }
```

5.5 调用唯一值渲染窗体

在【Theme】Tab 页上添加【Unique Value】按钮，建立 Click 响应函数。代码如下：

```
private void btnUniqueValue_Click(object sender,EventArgs e)
{
    UniqueValueRendererFrm frm=new UniqueValueRendererFrm ( m_map-
Contrl );
    If( frm.ShowDialog( )= =DialogRezult.OK )
    {
        _AxMapControl.ActiveView.Refresh();
        _AxMapControl.Update();
    }
}
```

5.6 编译运行

按下 F5 键，编译运行程序。

以上适用于 Windows 10+VS2015+AE10. 5 编译环境。

第6章　分级符号渲染

6.1　知识要点

分级渲染绘制方法是：先对要素图层的要素类中的字段值进行分级，然后对每个分级区间进行颜色渲染，ClassBreaksRendererClass 对象类实现了 IClassBreaksRendererClass 接口，该接口定义了几个重要属性和方法。

①Field：分级字段名；

②BreakCount：分级级数；

③set_Break(int Index，double Value)：设置分级对应的值；

④set_Symbol(int Index，ISymbol sym)：设置该分级的对应符号。

分级渲染绘制的步骤：

①遍历图层中的所有要素，获取渲染字段分级数组；

②创建一个 ClassBreaksRendererClass 对象；

③为每个分级区间配置一个相应符号；可按符号尺寸配置或按符号颜色配置，或两者都有；

④然后将渲染器赋给图层的 Renderer 属性。

6.2　功能描述

点击【Theme】Tab 页上【Class Symbol】按钮，弹出“分级符号渲染”对话框，如图 6-1 所示，可选择图层和符号化字段：这里设计支持三种方式：

①颜色变化，符号大小不变，样式字符串=Color；

②符号大小变化，颜色不变，样式字符串=Size；

③符号颜色和大小都变化，样式字符串=Both。

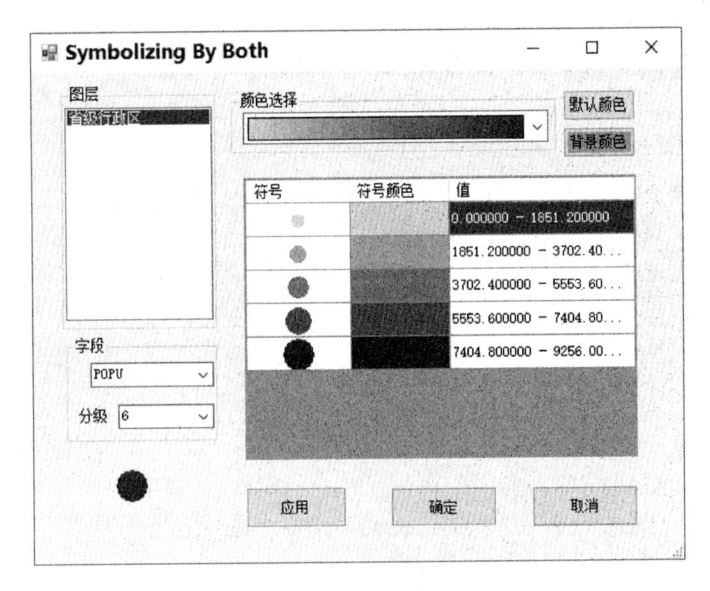

图 6-1　“分级符号渲染”对话框

6.3　功能实现

1. 功能类设计

新建一个 Windows 窗体，命名为“GraduatedSymbolsFrm. cs”。

从工具箱拖两个 ComBox(图层列表，选择字段)、两个 Button(btnOK、Cancel)控件到窗体，见表 6-1。

表 6-1　　控件及其属性

控　　件	Name 属性	Text 属性	其　　他
ListBox	ListBoxLayers	选择图层	
Combox	cbxSelField	选择字段	
Combox	cbxClassNumber	分级数量	
ComboxEx	imgcbxColorRamp	色带选择	自定义派生类
Button	btnBackgraoudColor	背景颜色	面类地物有用
Button	btnSymbolClolor	符号颜色	默认颜色
DataGridView	dataGridView1	显示值符号对	其中【符号】列类型为：DataGridViewImageColumn
Button	btnApp	应用	
Button	btnOK	确定	DialogRezult. OK
Button	btnCancel	取消	

添加代码如下：

```
public partial class GraduatedSymbolsFrm:Form
{
    IMapControl3 m_mapControl=null;
    private string m_strStyle="";
    //样式字符串属性(Size,Color,Both 之一)
    public string _StyleString
    {
        get  {  return m_strStyle; }
        set
        {
            m_strStyle=value;
            this.Text="Symbolizing By "+m_strStyle;
        }
    }

    public GraduatedSymbolsFrm(IMapControl3 mapControl)
    {
        InitializeComponent();
        m_mapControl=mapControl;

        //设置样式默认值：
        _StyleString="Size";
    }

    //图层选择响应函数：
    private void ListBoxLayers_SelectedIndexChanged(object sender,
EventArgs e)
    //字段选择响应函数：
    private void cbxSelFields_SelectedIndexChanged(object sender,
EventArgs e)
    //色带选择响应函数：
    private void imgcbxColorRamp_SelectedIndexChanged(object send-
er,EventArgs e)
    //分级选择响应函数：
    private void cbxClassNumber_SelectedIndexChanged(object sender,
EventArgs e)
    //确定按钮响应函数：
```

```
    private void btnOK_Click(object sender,EventArgs e)
    //应用按钮响应函数:
    private void btnApply_Click(object sender,EventArgs e)

    //核心函数:
    //获取分级数组
    public double[] GetClassesArray(IFeatureLayer pFeatLyr,string
sFieldName,int numclasses)
    //渲染调度函数
    public void GraduatedSymbolsRenderer(IFeatureLayer pFeatLyr,
string sFieldName,int numclasses)

    //若干辅助函数
    ……
}
```

2. 实现响应函数

代码如下:

```
//窗体加载时,填充图层列表框,初始化色带 ComboxEx
private void frmGraduatedSymbols_Load(object sender,EventArgs e)
{
    IEnumLayer layers=GetLayers();
    layers.Reset();
    ILayer layer=null;
    while ((layer=layers.Next())!=null)
    {
        ListBoxLayers.Items.Add(layer.Name);
    }

    //设置色带
    string sStyleFile=ESRI.ArcGIS.RuntimeManager.ActiveRuntime.Path;
    sStyleFile +="Styles\\ESRI.ServerStyle";
    InitColorRamp( sStyleFile);
    if (ListBoxLayers.Items.Count > 0)
    {
        this.ListBoxLayers.SelectedIndex=0;
        this.cbxClassNumber.SelectedIndex=5;
        this.imgcbxColorRamp.SelectedIndex=0;
```

```
        }
    }

    //图层选定后,响应函数将该层适合唯一值符号化的字段名,填充到字段下拉框
    private void ListBoxLayers_SelectedIndexChanged(object sender, EventArgs e)
    {
        IField pField=null;
        cbxSelFields.Items.Clear();
        cbxSelFields.Text="";

        IFeatureLayer pFtLayer=GetFeatureLayer(ListBoxLayers.SelectedItem.ToString());
        IFeatureClass pFeatCls=pFtLayer.FeatureClass;
        for (int i=0; i < pFeatCls.Fields.FieldCount; i++)
        {
            pField=pFeatCls.Fields.get_Field(i);
            if (pField.Type==esriFieldType.esriFieldTypeDouble ||
                pField.Type==esriFieldType.esriFieldTypeInteger ||
                pField.Type==esriFieldType.esriFieldTypeSingle ||
                pField.Type==esriFieldType.esriFieldTypeSmallInteger)
            {
                if (!cbxSelFields.Items.Contains(pField.Name))
                {
                    cbxSelFields.Items.Add(pField.Name);
                }
            }
        }
    }

    //字段选择响应函数,显示字段分级图(用到 GetClassesArray(),DisplayValues()),同时激活色带选择响应函数
    private void cbxSelFields_SelectedIndexChanged(object sender,EventArgs e)
    {
        if (this.cbxSelFields.SelectedIndex < 0)
            return;
        //获取字段的分级数组
```

```
        IFeatureLayer pFeatLyr=GetFeatureLayer(this.ListBoxLayers.SelectedItem.ToString());
        string sFieldName=this.cbxSelFields.SelectedItem.ToString();
        int numclasses=this.cbxClassNumber.SelectedIndex;
        double[] clsValues = GetClassesArray(pFeatLyr,sFieldName,numclasses);

        //重新填充值列
        DisplayValues(clsValues);
        //配置颜色符号
        imgcbxColorRamp_SelectedIndexChanged(sender,e);
        //应用按钮有效
        btnApply.Enabled=true;
    }
```

//遍历 DataGridView 中【值】列:
①对每列生成一个符号;符号的大小和颜色与_StyleString 相关;
②将 DataGridView 中【颜色】列的单元背景色设置为符号颜色;
③将 DataGridView 中【符号】列的单元值设置为依符号绘制的 Image;

```
    private void imgcbxColorRamp_SelectedIndexChanged(object sender, EventArgs e)
    {
        if (imgcbxColorRamp.SelectedIndex < 0)
            return;

        IFeatureLayer pFeatLyr=GetFeatureLayer(ListBoxLayers.SelectedItem.ToString());
        int ClassesCount=this.cbxClassNumber.SelectedIndex;

        //创建渐变色带
        IEnumColors pEnumColors=CreateAlgorithmicColorRamp(ClassesCount);
        //设置符号原始尺寸和颜色
        double OriginSize=(ClassesCount <=5) ? 8:7;
        IColor OriginColor = this.ColorToIRgbColor(this.btnDefaultColor.BackColor);

        //获取要素类的几何类型
```

```
        esriGeometryType shpType=pFeatLyr.FeatureClass.ShapeType;
        //需要注意的是,分级着色对象中的 symbol 和 break 的下标都是从 0 开始
        for (int i=0; i < DataGridView.RowCount; i++)
        {
            double nSize=OriginSize;
            if (_StyleString=="Size" ||_StyleString=="Both")
                nSize +=i*OriginSize/3.0d;

            IColor pNextColor=OriginColor;
            if (_StyleString=="Color" ||_StyleString=="Both")
                pNextColor=pEnumColors.Next();

            //生成不同的分级符号
            ISymbol pSymbol=CreateDefinedSymbol(shpType,nSize,pNextColor);

            //符号画成图片
            Size size=this.DataGridView[0,i].Size;
            Bitmap bitmap=SymbolToBitmap(pSymbol,0,size.Width,size.Height);
            this.pictureBox1.Image=bitmap;

            //填充 DataGridCell
            this.DataGridView[0,i].Value=bitmap;
            this.DataGridView[1,i].Style.BackColor = ColorTranslator.
FromOle(pNextColor.RGB);
        }
    }

    //分级选择响应函数:
    private void cbxClassNumber_SelectedIndexChanged(object sender,
EventArgs e)
    {
        cbxSelFields_SelectedIndexChanged(sender,e);
    }
    //OK,对选定图层和字段执行 GraduatedSymbolsRenderer(...)函数
    private void btnOK_Click(object sender,EventArgs e)
    {
        IFeatureLayer pFeatLyr=GetFeatureLayer(ListBoxLayers.Selecte-
dItem.ToString());
```

```
        string sFieldName=cbxSelFields.SelectedItem.ToString();
        int numClasses=int.Parse(this.cbxClassNumber.SelectedItem.To-
String());

        GraduatedSymbolsRenderer(pFeatLyr,sFieldName,numClasses);
        Close();
    }
```

3. 核心函数实现

GetClassesArray(...)是分级符号渲染核心函数，方法是先用 IBasicHistogram 接口获取渲染字段的出现过的值及其频数，然后使用 IClassifyGEN 进行等级划分(为简化起见，函数中仅用等间距分类法进行划分，ArcEngine 还支持自然断点划分、自定义划分、分位数划分、标准差划分等，可参考重分类)，得到分级数组。代码如下：

```
    public double[] GetClassesArray(IFeatureLayer pFeatLyr, string
sFieldName,int numclasses)
    {
        //获得要着色的图层
        IGeoFeatureLayer pGeoFeatureL=pFeatLyr as IGeoFeatureLayer;
        ITable pTable=pGeoFeatureL.FeatureClass as ITable;

        //创建哈希表
        ITableHistogram pTableHistogram=new BasicTableHistogramClass();
        IBasicHistogram pBasicHistogram=(IBasicHistogram)pTableHisto-
gram;
        pTableHistogram.Field=sFieldName;
        pTableHistogram.Table=pTable;

        //获取渲染字段的出现过值及其频数
        object dataValues;
        object dataFrequency;
        pBasicHistogram.GetHistogram(out dataValues,out dataFrequency);

        //依据"值-频数"数据进行等级划分
        IClassifyGEN pClassify=new EqualIntervalClass();
        try
        {
            pClassify.Classify(dataValues,dataFrequency,ref numclasses);
        }
        catch (Exception ex)
```

```
        {
        }
        //返回一个数组
        double[] dClasses=(double[])pClassify.ClassBreaks;
        return dClasses;
    }
```

GraduatedSymbolsRenderer(...)是分级渲染的调度函数：

①首先根据图层中选定字段获取分级数组；

②创建渐变色带；

③创建分级符号渲染器 GraduatedSymbolsRenderer；

④然后为不同分级配置不同大小和颜色的符号；

⑤渲染器绑定到图层。

这里用到几个辅助函数：

①GetClassesArray()：获取分级数组；

②CreateDefinedSymbol()：根据指定大小和颜色创建相应类型符号。

代码如下：

```
    public void GraduatedSymbolsRenderer(IFeatureLayer pFeatLyr, string sFieldName,int numclasses)
    {
        if (!check()) return;

        //获取分级数组
        double[] Classes = GetClassesArray(pFeatLyr, sFieldName, numclasses);
        int ClassesCount=Classes.GetUpperBound(0);

        //创建渐变色带
        IEnumColors pEnumColors=CreateAlgorithmicColorRamp(ClassesCount);

        //创建分级渲染器
        IClassBreaksRenderer pClassBreakRenderer = new ClassBreaksRendererClass();
        {
            pClassBreakRenderer.Field=sFieldName; //设置分级字段
            pClassBreakRenderer.BreakCount=ClassesCount;//设置分级数目
            pClassBreakRenderer.SortClassesAscending=true;//升序排列
```

```
        //设置背景颜色,面状地物需要
        IFillSymbol backgroundSymbol=new SimpleFillSymbolClass();
        backgroundSymbol.Color = ColorToIRgbColor(this.btnBackgroud-
Color.BackColor);
        pClassBreakRenderer.BackgroundSymbol=backgroundSymbol;

        //设置符号原始尺寸和颜色
        double OriginSize=(ClassesCount <=5) ? 8:7;
        IColor OriginColor=this.ColorToIRgbColor(this.btnDefault-
Color.BackColor);

        //获取要素类的几何类型
        esriGeometryType shpType=pFeatLyr.FeatureClass.ShapeType;
        //需要注意的是,分级着色对象中的 symbol 和 break 的下标都是从 0 开始
        for (int IbreakIndex = 0; IbreakIndex < ClassesCount; Ibrea-
kIndex++)
        {
            double nSize=OriginSize;
            if (_StyleString=="Size" ||_StyleString=="Both")
                nSize +=IbreakIndex*OriginSize/3.0d;

            IColor pNextColor=OriginColor;
            if (_StyleString=="Color" ||_StyleString=="Both")
                pNextColor=pEnumColors.Next();

            //不同的要素类型生成不同的分级符号
            ISymbol pSymbol =CreateDefinedSymbol(shpType,nSize,pNext-
Color);

            pClassBreakRenderer.set_Break(IbreakIndex,Classes[Ibrea-
kIndex+1]);
            pClassBreakRenderer.set_Symbol(IbreakIndex,pSymbol);
        }
    }

    //渲染器绑定到图层
    IGeoFeatureLayer pGeoFeatLyr=pFeatLyr as IGeoFeatureLayer;
    pGeoFeatLyr.Renderer=pClassBreakRenderer as IFeatureRenderer;
```

```
    }
```

4. 辅助函数

代码如下：

```
    //创建特定符号
    private ISymbol CreateDefinedSymbol ( esriGeometryType shpType,
double nSize,IColor pNextColor)
    {
        ISymbol pISymbol=null;
        switch (shpType)
        {
            case esriGeometryType.esriGeometryPolygon:
                {
                    IMarkerSymbol pMarkerSymbol=new SimpleMarkerSymbolClass
();
                    pMarkerSymbol.Color=pNextColor;
                    pMarkerSymbol.Size=nSize;

                    pISymbol=pMarkerSymbol as ISymbol;
                    break;
                }
            case esriGeometryType.esriGeometryPolyline:
                {
                    ILineSymbol pLineSymbol=new SimpleLineSymbolClass();
                    pLineSymbol.Color=pNextColor;
                    pLineSymbol.Width=nSize;

                    pISymbol=pLineSymbol as ISymbol;
                    break;
                }
            case esriGeometryType.esriGeometryPoint:
                {
                    IMarkerSymbol pMarkerSymbol=new SimpleMarkerSymbolClass
();
                    pMarkerSymbol.Color=pNextColor;
                    pMarkerSymbol.Size=nSize;

                    pISymbol=pMarkerSymbol as ISymbol;
                    break;
```

```
            }
        }

        return pISymbol;
    }
    //创建渐变色带
    private IEnumColors CreateAlgorithmicColorRamp(int ClassesCount)
    {
        if (imgcbxColorRamp.SelectedIndex < 0)
            return null;

        IColor fromColor=((ItemEx)imgcbxColorRamp.SelectedItem).From-
Color;
        IColor toColor=((ItemEx)imgcbxColorRamp.SelectedItem).ToColor;

        //建立从黄到红的渐变色
        IAlgorithmicColorRamp pRamp=new AlgorithmicColorRamp();
        pRamp.Algorithm=ESRI.ArcGIS.Display.esriColorRampAlgorithm.
esriHSVAlgorithm;
        pRamp.FromColor=fromColor;
        pRamp.ToColor=toColor;
        pRamp.Size=(ClassesCount >5 )? ClassesCount:5;

        bool ok=false;
        pRamp.CreateRamp(out ok);
        return pRamp.Colors;
    }
    //DataGridView 上显示分级区间
    private void DisplayValues(double[] clsValues)
    {
        DataGridView.Rows.Clear();
        for (int i=0; i < clsValues.Length - 1; i++)
        {
            //将属性值添加到表格
            object vntUniqueValue=clsValues[i].ToString("F6")+" - " +
                                  clsValues[i+1].ToString("F6");;
            {
                DataGridView.Rows.Add();
```

```
                DataGridView[2,i].Value=vntUniqueValue;
            }
        }
    }
```

此外，还涉及两个 GetFeatureLayer(...)和 GetLayers(...)，ColorToIRgbColor(...)，InitColorRamp(...)辅助函数，请参阅之前有关内容。

另外，用到 SymbolToBitmap()内联函数将符号画成 Image 对象，以及两个辅助函数：

```
    internal System.Drawing.Bitmap SymbolToBitmap(ISymbol iSymbol,int
iStyle,int iWidth,int iHeight)
    {
        //设置 tag 矩形
        tagRECT itagRECT=default(ESRI.ArcGIS.esriSystem.tagRECT);
        {
            itagRECT.left=0;
            itagRECT.right=iWidth;
            itagRECT.top=0;
            itagRECT.bottom=iHeight;
        }
        //设置 Ersi 矩形;
        IEnvelope iEnvelope=new EnvelopeClass();
        iEnvelope.PutCoords(0,0,iWidth,iHeight);
        //创建画板
        Bitmap iBitmap=new System.Drawing.Bitmap(iWidth,iHeight);
        Graphics iGraphics=System.Drawing.Graphics.FromImage(iBitmap);

        //创建显示转换器
        IDisplayTransformation iDisplayTransformation = new DisplayTrans-
formationClass();
        {
            //设置范围
            iDisplayTransformation.VisibleBounds=iEnvelope;
            iDisplayTransformation.Bounds=iEnvelope;
            iDisplayTransformation.set_DeviceFrame(ref itagRECT);

            //设置分辨率
            iDisplayTransformation.Resolution=iGraphics.DpiX;
        }
```

```
        IntPtr iHDC=new IntPtr();
        try
        {
            iHDC=iGraphics.GetHdc();
        }
        catch (Exception ex)
        {
            throw new Exception(string.Format("Raster::Symbolize: \r\n
{0}",ex.Message));
        }

        //创建几何对象
        IGeometry iGeometry=ConstructGeometry( iSymbol,iStyle,iWidth,
iHeight);

        //绘制图形
        iSymbol.SetupDC(iHDC.ToInt32(),iDisplayTransformation);
        iSymbol.Draw(iGeometry);
        iSymbol.ResetDC();
        iGraphics.ReleaseHdc(iHDC);
        iGraphics.Dispose();
        return iBitmap;
    }
    //构造一个几何对象
    internal IGeometry ConstructGeometry(ISymbol iSymbol, int iStyle,
int iWidth,int iHeight)
    {
        IPoint iPoint=default(IPoint);
        IGeometryCollection iPolyline=default(IGeometryCollection);
        IGeometryCollection iPolygon=default(IGeometryCollection);
        IRing iRing=default(IRing);
        ISegmentCollection iSegmentCollection=default(ISegmentCollec-
tion);
        IGeometry iGeometry=default(IGeometry);
        object Missing=Type.Missing;

        //获取 Geometry
        if ((iSymbol) is IMarkerSymbol)
```

```
        {
            switch (iStyle)
            {
                case 0:
                    iPoint=new ESRI.ArcGIS.Geometry.Point();
                    iPoint.PutCoords(iWidth/2,iHeight/2);
                    iGeometry=iPoint;
                    break;
            }
        }
        else if ((iSymbol) is ILineSymbol)
        {
            iSegmentCollection=new ESRI.ArcGIS.Geometry.Path() as ISeg-
mentCollection;
            iPolyline=new Polyline() as IGeometryCollection;
            switch (iStyle)
            {
                case 0:
                    iSegmentCollection.AddSegment(CreateLine(0,iHeight/2,
iWidth,iHeight/2),ref Missing,ref Missing);
                    iPolyline.AddGeometry(iSegmentCollection as IGeome-
try,ref Missing,ref Missing);
                    iGeometry=iPolyline as IGeometry;
                    break;
                case 1:
                    iSegmentCollection.AddSegment(CreateLine(0,iHeight/4,
iWidth/4,3*iHeight/4),ref Missing,ref Missing);
                    iSegmentCollection.AddSegment(CreateLine(iWidth/4,
3*iHeight/4,3*iWidth/4,iHeight/4),ref Missing,ref Missing);
                    iSegmentCollection.AddSegment(CreateLine(3*iWidth/
4,iHeight/4,iWidth,3*iHeight/4),ref Missing,ref Missing);
                    iPolyline.AddGeometry(iSegmentCollection as IGeome-
try,ref Missing,ref Missing);
                    iGeometry=iPolyline as IGeometry;
                    break;
            }
        }
        else if ((iSymbol) is IFillSymbol)
```

```
        {
            iSegmentCollection=new Ring() as ISegmentCollection;
            iPolygon=new Polygon() as IGeometryCollection;
            switch (iStyle)
            {
                case 0:
                    iSegmentCollection.AddSegment(CreateLine(5,iHeight -2,
iWidth -5,iHeight -2),ref Missing,ref Missing);
                    iSegmentCollection.AddSegment(CreateLine(iWidth -5,
iHeight -2,iWidth -5,2),ref Missing,ref Missing);
                    iSegmentCollection.AddSegment(CreateLine(iWidth -5,
2,5,2) as ISegment,ref Missing,ref Missing);
                    iRing=iSegmentCollection as IRing;
                    iRing.Close();
                    iPolygon.AddGeometry(iSegmentCollection as IGeometry,
ref Missing,ref Missing);
                    iGeometry=iPolygon as IGeometry;
                    break;
            }
        }
        else if ((iSymbol) is ESRI.ArcGIS.Display.ISimpleTextSymbol)
        {
            switch (iStyle)
            {
                case 0:
                    iPoint=new ESRI.ArcGIS.Geometry.Point();
                    iPoint.PutCoords(iWidth/2,iHeight/2);
                    iGeometry=iPoint;
                    break;
            }
        }

        return iGeometry;
    }
    //创建一条线段
    internal ISegment CreateLine(double x1,double y1,double x2,double y2)
    {
        IPoint P1=new PointClass();
```

```
    P1.PutCoords(x1,y1);
    IPoint P2=new PointClass();
    P2.PutCoords(x2,y2);

    ILine ln=new LineClass();
    ln.PutCoords(P1,P2);
    return (ln as ISegment);
}
```

6.4 调用分级符号渲染窗体

在【Theme】Tab 页上添加【Class Symbol】按钮，建立 Click 响应函数，代码如下：

```
private void btnClassSymbol_Click(object sender,EventArgs e)
{
    GraduatedSymbolsFrm frm=new GraduatedSymbolsFrm( m_mapContrl );
    If( frm.ShowDialog( )= =DialogRezult.OK )
    {
        _AxMapControl.ActiveView.Refresh();
        _AxMapControl.Update();
    }
}
```

6.5 编译运行

按下 F5 键，编译运行程序。

第7章　统计图表符号渲染

7.1　知识要点

统计图表是专题地图中经常使用的一类符号，用来比较一个要素的多个属性的比率关系。常用的统计图表类型有饼状图、条形图、柱状图、堆叠图等。

在 ArcGIS Engine 中，无论是制作饼状图、条形图、柱状图还是堆叠图，都是由 ChartRenderer 类实现的，该类实现了 IChartRenderer，IRendererFields，IMarkerSymbol 等接口。IChartRenderer 接口主要属性如下：

①BaseSymbol 属性用于设置背景填充符号(当统计图绘制在面要素上时)。

②ChartSymbol 属性用于设置统计图表的样式，包括 IBarChartSymbol，IPieChartSymbol 和 IStackedChartSymbol 等。

③IChartSymbol 接口的 MaxValue 属性设置统计值中的最大值。

④IRendererFields 接口记录参与计算的字段集合。IMarkerSymbol 接口的 Size 属性设置统计图表形状的最大值，如在柱状图中最大值表示最大高度。

实例代码的实现思路如下：

①遍历所选每一个字段的属性值的最大值，记录属性集合。

②根据不同的选择生成对应的图表符号。如果是条形图，则生成 BarChartSymbol。

③创建相应图表渲染器，并对渲染器属性赋值，包括：MaxValue 属性、ChartSymbol 属性、BaseSymbol 属性、字段集合 IRendererFields 填充。

④将图表渲染对象与渲染图层挂钩。

7.2　功能描述

点击【Theme】Tab 页上【Chart Symbol】按钮，弹出“统计图表符号渲染”对话框，如图 7-1 所示，可选择若干字段参与符号化计算：

功能设计支持三种不同样式：

①饼状图：样式类型字符=Bie；

②柱状图：样式类型字符=Bar；

③堆叠图：样式类型字符=Stacked。

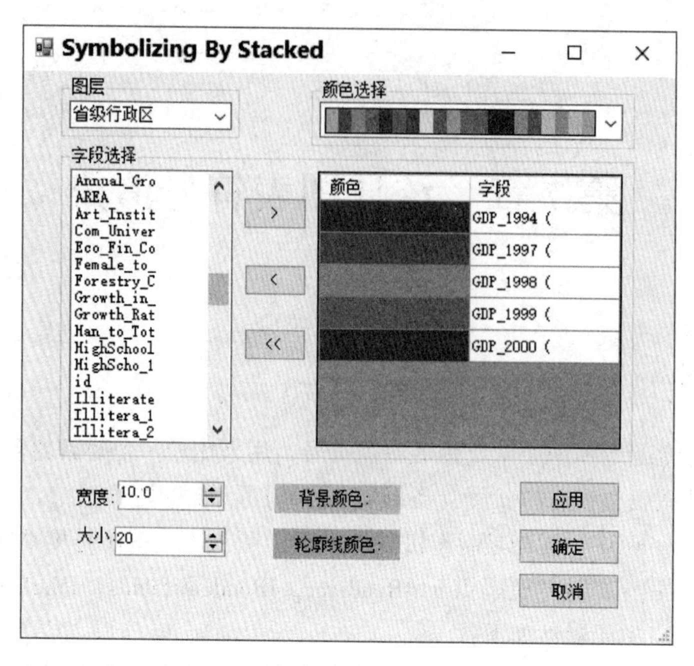

图 7-1 “统计图表符号渲染”对话框

7.3 功能实现

1. 功能类设计

新建一个 Windows 窗体，命名为“ChartSymbolFrm. cs”。

从工具箱拖动表 7-1 的控件到窗体。

表 7-1 控件及其属性

控　件	Name 属性	Text 属性	其　他
ListBox	listBoxFields	选择字段	
Combox	cbxLyrSelected	选择图层	
ComboxEx	imgcbxColorRamp	色带选择	自定义派生类
NumericUpDown	numUpDownWidth	符号宽度	柱状图，堆叠图用到
NumericUpDown	numUpDownSize	符号大小	柱状图，堆叠图时，代表其最大高度
Panel	PanelBackGroudColor	背景颜色	面类地物有用
Panel	panelLineColor	轮廓线颜色	默认颜色
DataGridView	dataGridView1	显示字段-符号对	
Button	btnSingleAdd	添加一个字段	

续表

控　　件	Name 属性	Text 属性	其　　他
Button	btnSingleRemove	移除一个字段	
Button	btnAllRemove	移除所有字段	
Button	btnApp	应用	
Button	btnOK	确定	DialogRezult. OK
Button	btnCancel	取消	

添加如下代码：

```
public partial class ChartSymbolFrm:Form
{
    private IMapControl3 m_mapControl;
    private DataTable m_pDataTable=null;
    private string m_strStyle="";
    //样式字符串(Pie,Bar,Stacked 之一)
    public string _StyleString
    {
        get  { return m_strStyle;   }
        set
        {
            m_strStyle=value;
            this.Text="Symbolizing By "+m_strStyle;
        }
    }
    public ChartSymbolFrm(IMapControl3 mapControl)
    {
        InitializeComponent();
        m_mapControl=mapControl;

        m_pDataTable=new DataTable();
        _StyleString="Pie"; //设置图表类型默认值
    }
    //事件响应函数
    private void ChartSymbolFrm_Load(object sender,EventArgs e)
    private void cbxLyrSelected_SelectedIndexChanged(object sender,
EventArgs e)
    private void imgcbxColorRamp_SelectedIndexChanged(object send-
```

```
er,EventArgs e)
        private void btnSingleAdd_Click(object sender,EventArgs e)
        private void btnSingleRemove_Click(object sender,EventArgs e)
        private void btnAllRemove_Click(object sender,EventArgs e)
        private void btnOk_Click(object sender,EventArgs e)
        private void btnApply_Click(object sender,EventArgs e)

        //核心函数
        private IChartSymbol CreateChartSymbol(string strType,double
maxValue,bool isDislay3D)
        //若干功能函数
        ……
    }
```

2. 实现响应函数

实现代码如下：

```
//装载事件响应函数
private void ChartSymbolFrm_Load(object sender,EventArgs e)
{
    //填充 DataTable 表头
    DataColumn symbolCol=new DataColumn("颜色");
    DataColumn valueCol=new DataColumn("字段");
    m_pDataTable.Columns.Add(symbolCol);
    m_pDataTable.Columns.Add(valueCol);

    //数据视图控件捆绑数据源
    dataGridView1.DataSource=m_pDataTable;
    dataGridView1.AllowUserToAddRows=false;
    btnAllRemove.Enabled=false;

    //填充图层下拉框
    IEnumLayer layers=GetLayers();
    layers.Reset();
    ILayer layer=null;
    while ((layer=layers.Next())!=null)
    {
        this.cbxLyrSelected.Items.Add(layer.Name);
    }
    //初始化色带下拉框
```

```
        string sStyleFile=ESRI.ArcGIS.RuntimeManager.ActiveRuntime.Path;
        sStyleFile  +="Styles\\ESRI.ServerStyle";
        InitColorRamp(sStyleFile);

        if ((cbxLyrSelected.Items.Count > 0))
        {
            cbxLyrSelected.SelectedIndex=0;
        }
    }
    //图层选定后,响应函数将该层合法字段名填充到字段列表框
    private void cbxLyrSelected_SelectedIndexChanged(object sender,
EventArgs e)
    {
        if (cbxLyrSelected.SelectedIndex==-1)
            return;
        //得到选中的图层的字段集合
        string lyrName=cbxLyrSelected.SelectedItem.ToString();
        IGeoFeatureLayer pGeoFeatureLayer=(IGeoFeatureLayer)GetFeatu-
reLayer(lyrName);
        IArray pLayerFieldsArray=GetLayerFields(pGeoFeatureLayer);
        if (pLayerFieldsArray==null)
            return;

        //填充字段列表框
        listBoxFields.Items.Clear();
        for (int i=0; i < pLayerFieldsArray.Count; i++)
        {
            //过滤字段
            IField pField=(IField)pLayerFieldsArray.get_Element(i);
            if ((!(pField.Name=="FID")) & (!(pField.Name=="ID"))
                & (pField.VarType==2 |pField.VarType==3 |
                    pField.VarType==4 |pField.VarType==5))
            {
                //Dim pClassify As IClassify
                ITableHistogram pTableHistogram=default(ITableHistogram);
                IBasicHistogram pHistogram=default(IBasicHistogram);
                object vntDataFrequency=null;
                object vntDataValues=null;
```

```
                pTableHistogram=new BasicTableHistogram() as ITableHis-
togram;
                pTableHistogram.Field=pField.Name;
                pTableHistogram.Table=pGeoFeatureLayer as ITable;

                pHistogram=pTableHistogram as IBasicHistogram;
                pHistogram.GetHistogram(out vntDataValues,out vntDataFre-
quency);

                //数据类型必须为"数据"
                string strDataType=null;
                strDataType=Information.TypeName(vntDataValues);

                //该字段符合分类渲染的要求,加入之
                if(strDataType=="Integer()" |strDataType=="Long()" |
strDataType=="Double()")
                    listBoxFields.Items.Add(pField.Name);
            }
        }
    }
    //遍历 DataGridView 中【颜色】列:将单元背景色设置为色带指定颜色
    private void imgcbxColorRamp_SelectedIndexChanged(object sender,
EventArgs e)
    {
        if (imgcbxColorRamp.SelectedIndex < 0)
            return;

        //将色带颜色值显示在 DataGridView 中
        IEnumColors pColorEnumerater = ((ItemEx) imgcbxColorRamp.Selecte-
dItem).ColorEnumerater;
        pColorEnumerater.Reset();
        for (int i=0; i < this.dataGridView1.RowCount; i++)
        {
            IColor pNextUniqueColor=pColorEnumerater.Next();
            if (pNextUniqueColor==null)
            {
                pColorEnumerater.Reset();
```

```
                pNextUniqueColor=pColorEnumerater.Next();
            }

            Color pColor=ColorTranslator.FromOle(pNextUniqueColor.RGB);
            this.dataGridView1[0,i].Style.BackColor=pColor;
        }
    }
//根据字段列表框中选定的字段,为 DataTable 添加新行
//更新数据表数据源
//移除字段列表框中被选字段
//激活色带选择事件重新配色
private void btnSingleAdd_Click(object sender,EventArgs e)
{
    if ((listBoxFields.SelectedIndex==-1))
    {
        MessageBox.Show("请选择要添加的字段!","提示");
        return;
    }
    btnAllRemove.Enabled=true;
    dataGridView1.AllowUserToAddRows=true;

    //向 m_pDataTable 表中添加一行
    DataRow pDatarow=m_pDataTable.NewRow();
    pDatarow[1]=listBoxFields.SelectedItem.ToString();
    m_pDataTable.Rows.Add(pDatarow);

    //更新 gridview 数据源
    dataGridView1.DataSource=m_pDataTable;

    //删除 Field 相关项
    listBoxFields.Items.RemoveAt(listBoxFields.SelectedIndex);

    //重新配色
    imgcbxColorRamp_SelectedIndexChanged(sender,e);
    dataGridView1.AllowUserToAddRows=false;
    this.btnApply.Enabled=true;
}
//数据表中选定一行移动到字段列表框
```

```
    private void btnSingleRemove_Click(object sender,EventArgs e)
    {
        if ((dataGridView1.SelectedCells.Count==0))
        {
            MessageBox.Show("请选择要移除的字段!","提示");
            return;
        }

        DataGridViewRow pRow = dataGridView1.Rows[dataGridView1.Cur-
rentCell.RowIndex];
        listBoxFields.Items.Add(pRow.Cells[1].Value.ToString());
        dataGridView1.Rows.RemoveAt(dataGridView1.CurrentCell.RowIndex);
        if ((this.dataGridView1.Rows.Count==0))
        {
            this.btnApply.Enabled=false;
        }
    }
    //将数据表所有字段移动到字段列表框
    private void btnAllRemove_Click(object sender,EventArgs e)
    {
        if ((dataGridView1.Rows.Count <=0))
        {
            MessageBox.Show("无可移植的字段!","提示",MessageBoxButtons.OK,
                                                MessageBoxIcon.Infor-
mation);
            return;
        }

        //FieldListBox 添加项
        for (int i=0; i <=dataGridView1.Rows.Count - 1; i++)
        {
            listBoxFields.Items.Add(dataGridView1.Rows[i].Cells[1].
Value.ToString());
        }

        //更新 gridview 数据源
        m_pDataTable.Clear();
        dataGridView1.DataSource=m_pDataTable;
```

```
        btnAllRemove.Enabled=false;
        btnAllAdd.Enabled=true;
        this.btnApply.Enabled=false;
    }
    //OK/App 响应函数
    private void btnOk_Click(object sender,EventArgs e)
    {
        if (btnApply.Enabled==true)
        {
            btnApply_Click(null,null);
        }
        this.Close();
    }
    private void btnApply_Click(object sender,EventArgs e)
    {
        ChartRenderer();
    }
```

3. 核心函数实现

ChartRenderer()函数是图表渲染总的调度函数，实现步骤如下：

①创建背景符号；

②创建图表符号；

③创建图表渲染对象；

④将图表渲染对象与渲染图层挂钩。

具体代码如下：

```
    private void ChartRenderer()
    {
         int FieldsCount=this.dataGridView1.Rows.Count;
        if (FieldsCount < 2)
            return;
        //获取图层接口
        IFeatureLayer _pLayer=GetFeatureLayer(this.cbxLyrSelected.Se-
lectedItem.ToString());

        //创建背景符号：
        ISymbol _pBaseSymbol=default(ISymbol);
        {
            //生成背景颜色
```

```
            IColor BackGroudColor = ColorToIRgbColor(PanelBackGroudColor.
BackColor);
            BackGroudColor.UseWindowsDithering=true;
            _pBaseSymbol= CreateDefinedSymbol (_ pLayer. FeatureClass.
ShapeType,BackGroudColor);
        }

        //创建图表符号
        double maxValue=GetMaxMinValue((ITable)_pLayer,false); //获取
最大值
        IChartSymbol_pChartSymbol=CreateChartSymbol(_StyleString,maxVal-
ue,true);

        //创建图表渲染对象
        IChartRenderer pChartRenderer=new ChartRenderer();
        {
            //向渲染字段对象中添加选中的字段
            IRendererFields pRendererFields = (IRendererFields) pChar-
tRenderer;
            for (int i=0; i < FieldsCount; i++)
            {
                string strValue=dataGridView1[1,i].Value.ToString();
                pRendererFields.AddField(strValue,strValue);
            }

            //设置背景
            pChartRenderer.BaseSymbol=_pBaseSymbol;
            pChartRenderer.UseOverposter=false;

            //图表符号赋值给渲染器
            pChartRenderer.ChartSymbol=_pChartSymbol;
            pChartRenderer.CreateLegend();

            //使用IPieChartRenderer接口设置有关属性
            IPieChartRenderer pPieChartRender=pChartRenderer as IPieC-
hartRenderer;
            {
                ////按照比例显示
```

```
                //pPieChartRender.ProportionalBySum=false;

                //字段的最大值和最小值
                pPieChartRender.MinSize=30;
                pPieChartRender.MinValue=0;
                pPieChartRender.FlanneryCompensation=false;
            }
        }

        //将图表渲染对象与渲染图层挂钩
        IGeoFeatureLayer pGeoFeatureLayer=(IGeoFeatureLayer)_pLayer;
        pGeoFeatureLayer.Renderer=(IFeatureRenderer)pChartRenderer;
        this.btnOk.Enabled=true;
    }

    //CreateChartSymbol()函数根据类型字符串,创建不同类型的图表符号
    private IChartSymbol CreateChartSymbol(string strType,double max-
Value,bool isDislay3D)
    {
        IChartSymbol pChartSymbol=null;
        switch (strType)
        {
            case "Pie":
                //创建饼图符号
                 IPieChartSymbol pPieChartSymbol =new PieChartSymbolClass
();
                {
                    pPieChartSymbol.Clockwise=true;

                    //设置轮廓线
                    ILineSymbol pOutLine=new SimpleLineSymbolClass();
                    pOutLine.Color=ColorToIRgbColor(panelLineColor.Back-
Color);
                    pOutLine.Width=1.0;
                    pPieChartSymbol.UseOutline=true;
                    pPieChartSymbol.Outline=pOutLine;

                    //设置饼状图符号大小
```

```
                    IMarkerSymbol pMarkerSymbol=(IMarkerSymbol)pPieCh-
artSymbol;
                    pMarkerSymbol.Size = Convert.ToDouble(numUpDownSize.
Value);

                    //填充符号数组
                    ISymbolArray pSymbolArray=(ISymbolArray)pPieChart-
Symbol;
                    FillSymbolArrayWithColor(pSymbolArray);
                }
                pChartSymbol=(pPieChartSymbol as IChartSymbol);
                break;
            case "Bar":
                IBarChartSymbol pBarChartSymbol=new BarChartSymbolClass();
                {
                    //设置柱状图宽度
                    pBarChartSymbol.Width=Convert.ToDouble(numUpDown-
Width.Value);

                    //设置柱状图最大高度
                    IMarkerSymbol pMarkerSymbol=(IMarkerSymbol)pBarCh-
artSymbol;
                    pMarkerSymbol.Size = Convert.ToDouble(numUpDownSize.
Value);

                    //填充符号数组
                    ISymbolArray pSymbolArray=(ISymbolArray)pBarChart-
Symbol;
                    FillSymbolArrayWithColor(pSymbolArray);
                }
                pChartSymbol=(pBarChartSymbol as IChartSymbol);
                break;
            case "Stacked":
                IStackedChartSymbol pStackedChartSymbol=new StackedCh-
artSymbolClass();
                {
                    //设置轮廓线
                    ILineSymbol pOutLine=new SimpleLineSymbolClass();
```

```
                pOutLine. Color = ColorToIRgbColor ( panelLineColor.
BackColor);
                pOutLine.Width=1.0;
                pStackedChartSymbol.UseOutline=true;
                pStackedChartSymbol.Outline=pOutLine;

                //设置堆状图宽度
                pStackedChartSymbol.Width=Convert.ToDouble(numUp-
DownWidth.Value);

                //设置堆状图最大高度
                IMarkerSymbol pMarkerSymbol = (IMarkerSymbol) pStacked-
ChartSymbol;
                pMarkerSymbol. Size = Convert. ToDouble ( numUpDownSize.
Value);

                //填充符号数组
                ISymbolArray pSymbolArray=(ISymbolArray)pStackedChart-
Symbol;
                FillSymbolArrayWithColor(pSymbolArray);
            }
            pChartSymbol=(pStackedChartSymbol as IChartSymbol);
            break;
    }

    pChartSymbol.MaxValue=maxValue;
    I3DChartSymbol p3DChartSymbol=(pChartSymbol as I3DChartSymbol);
    p3DChartSymbol.Display3D=isDislay3D;

    return pChartSymbol;
}
```

4. 辅助函数

代码如下：

```
private double GetMaxMinValue(ITable pTable,bool isMinimum)
{
    double minValue=double.MaxValue;
    double maxValue=double.MinValue;
```

```
        int FieldsCount=this.dataGridView1.Rows.Count;
        for (int i=0; i < FieldsCount; i++)
        {
            ICursor pCursor=pTable.Search(null,false);
            string FieldName=dataGridView1[1,i].Value.ToString();

            IDataStatistics pDataStaticstics=new DataStatistics();
            pDataStaticstics.Cursor=pCursor;
            pDataStaticstics.Field=FieldName;

            IStatisticsResults pStatisticsResults = pDataStaticstics.Sta-
tistics;
            double value=Convert.ToDouble(pStatisticsResults.Maximum);

            if (value > maxValue)
                maxValue=value;

            if (value < minValue)
                minValue=value;
        }

        if (minValue <=0)
        {
            MessageBox.Show("最小值是零或小于零","Message",MessageBox-
Buttons.OK,MessageBoxIcon.Information);
        }
        return (isMinimum ? minValue:maxValue);
    }

    public IArray GetLayerFields(IFeatureLayer pfeaturelayer)
    {
        IGeoFeatureLayer pFeaturelyr=(IGeoFeatureLayer)pfeaturelayer;
        IFeatureClass pFeatureClass=pFeaturelyr.DisplayFeatureClass;
        if (pFeatureClass==null)
            return null;

        IArray pAryLayerField=new ESRI.ArcGIS.esriSystem.Array();//图
层字段队列接口
```

```
        try
        {
            IFields pFields=pFeatureClass.Fields;
            for (int intFieldIndex=0; intFieldIndex<pFields.FieldCount;
intFieldIndex++)
            {
                IField pField=pFields.get_Field(intFieldIndex);
                if (pField.Name.ToUpper()!="SHAPE" & pField.Name.ToUpper
()!="SHAPE.LEN")
                {
                    pAryLayerField.Add(pField);
                }
            }
        }
        catch (Exception ex)
        {
            MessageBox.Show(ex.ToString());
        }

        return pAryLayerField;
    }

    private void FillSymbolArrayWithColor(ISymbolArray pSymbolsArray)
    {
        int FieldsCount=this.dataGridView1.Rows.Count;
        //当前渲染字段的个数
        if (FieldsCount==0)
            return;

        for (int i=0; i < FieldsCount; i++)
        {
            Color  color=dataGridView1[0,i].Style.BackColor;
            IColor pColor=this.ColorToIRgbColor(color);
            SimpleFillSymbol pFillSymbol=new SimpleFillSymbol();

            pFillSymbol.Color=pColor;
            pFillSymbol.Style=esriSimpleFillStyle.esriSFSSolid;
            pSymbolsArray.AddSymbol(pFillSymbol as ISymbol);
```

```
    }
}
```

这里只列出 GetMaxMinValue( )，GetLayerFields( )，FillSymbolArrayWithColor( )函数。还涉及两个 GetFeatureLayer(...)和 GetLayers(...)，ColorToIRgbColor(...)，CreateDefinedSymbol( )，InitColorRamp(...)辅助函数，请参阅之前有关内容。

7.4 调用分级符号渲染窗体

在【Theme】Tab 页上添加【Chart Symbol】按钮，建立响应函数，代码如下：

```
private void btnChartSymbol_Click(object sender,EventArgs e)
{
    ChartSymbolFrm frm=new ChartSymbolFrm(m_mapControl);
    frm._StyleString="Stacked";//"Bar";
    if (frm.ShowDialog()= =DialogResult.OK)
    {
        //更新主 Map 控件和图层控件
        this._AxMapControl.ActiveView.Refresh();
        this._AxMapControl.Update();
    }
}
```

7.5 编译运行

按下 F5 键，编译运行程序。

第8章　栅格数据渲染

8.1　知识要点

栅格数据渲染主要有两种方法：一是唯一值色彩渲染法，适用于离散有限值域栅格数据，用到 UniqueValueRendererClass 类，用法与要素类唯一值渲染类似；二是分级色彩渲染法，用到 IRasterClassifyColorRampRenderer 类，用法与要素类分级渲染染类似。它们与要素类的主要区别在于：取得唯一值集合和分级数组的方法不同。

对于唯一值色彩渲染法，栅格数据的属性表已经包含了唯一值集合，比矢量数据更简单。

对于分级色彩渲染法，如果栅格数据存在属性表（或能够重建属性表），属性表已包含值（Value）-频数（Count）数据，可以使用 Histogram 提取 Value-Frequency 数组。如果不存在属性表，使用波段（通常是第一波段）的 Histogram 表获取波段 0~255 值的频数数组，然后将 0~255 值拉伸（或压缩）为实际值数组。

分级渲染绘制的步骤：

①获取分级数组；

②创建一个 IRasterClassifyColorRampRenderer 对象；

③为每个分级区间配置一个相应颜色；

④最后将渲染器赋给图层的 Renderer 属性。

8.2　功能描述

点击【Theme】Tab 页上【Raster Graduate】按钮，弹出“栅格分级渲染”对话框，如图 8-1 所示，可选择图层和符号化字段。

点击“确定”按钮，得到符号化效果。

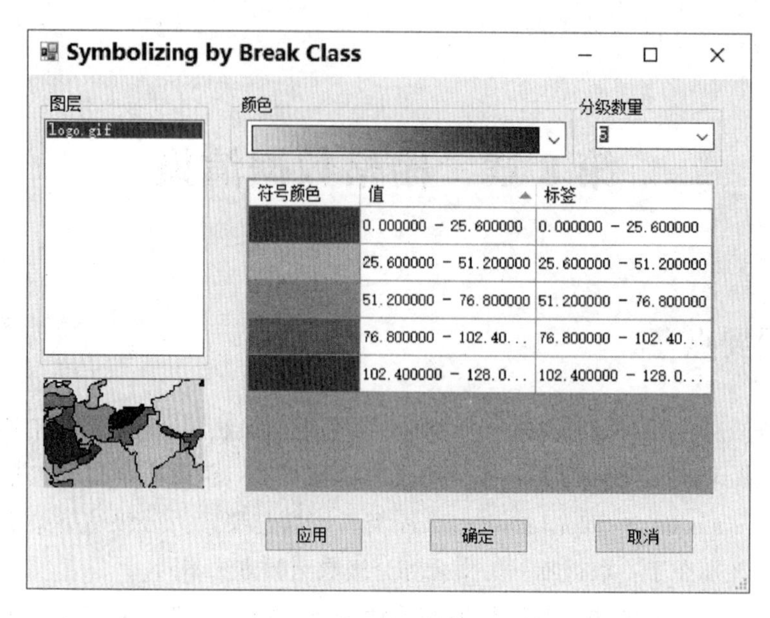

图 8-1 "栅格分级渲染"对话框

8.3 功能实现

1. 功能类设计

新建一个 Windows 窗体，命名为"RasterGraduateSymbolizeFrm. cs"。

从工具箱拖动表 8-1 中的控件到窗体。

表 8-1 控件及其属性

控 件	Name 属性	Text 属性	其 他
ListBox	ListBoxLayers	选择图层	
Combox	cbxClassNumber	分级数量	
ComboxEx	imgcbxColorRamp	色带选择	自定义派生类
DataGridView	dataGridView1	显示值符号对	
Button	btnApp	应用	
Button	btnOK	确定	DialogRezult. OK
Button	btnCancel	取消	

添加如下代码：

```
public partial class RasterGraduateSymbolizeFrm:Form
```

```
{
    private IMapControl3 m_mapControl=null;
    private IRasterBand m_bandRaster=null;
    private bool _isTableExisted=true;
    public RasterGraduateSymbolizeFrm(IMapControl3 mapControl)
    {
        InitializeComponent();
        m_mapControl=mapControl;
    }
    //事件响应函数
    private void RasterGraduateSymbolizeFrm_Load(object sender,
EventArgs e)
    private void ListBoxLayers_SelectedIndexChanged(object sender,
EventArgs e)
    private void cbxClassNumber_SelectedIndexChanged(object sender,
EventArgs e)
    private void imgcbxColorRamp_SelectedIndexChanged(object send-
er,EventArgs e)
    private void btnOk_Click(object sender,EventArgs e)
    private void btnApply_Click(object sender,EventArgs e)
    private void btnCancel_Click(object sender,EventArgs e)

    //核心功能函数
    private void RasterBreakClassRenderer()
    private double[] CreateStretchBreakClass(IRasterBand pRsBand,
int numDesiredClasses)
    private double[] CreateBreakClass(IRasterBand pBand,int num-
DesiredClasses)

    //若干辅助函数
    private bool RebuildRasterAttribute(IWorkspace workspace,
string rstFileName)
    ……
}
```

2. 实现响应函数

(1)加载响应函数:填充图层列表框,初始化色带 ComboxEx,代码如下:

```
private void RasterGraduateSymbolizeFrm_Load(object sender,Event-
Args e)
```

```
{
    IEnumLayer layers=GetRasterLayers();
    layers.Reset();
    ILayer layer=null;
    while ((layer=layers.Next())!=null)
    {
        ListBoxLayers.Items.Add(layer.Name);
    }

    //设置色带
    string sStyleFile=ESRI.ArcGIS.RuntimeManager.ActiveRuntime.Path;
    sStyleFile +="Styles\\ESRI.ServerStyle";
    InitColorRamp(sStyleFile);

    this.cbxClassNumber.SelectedIndex= 5;
    if (ListBoxLayers.Items.Count > 0)
    {
        ListBoxLayers.SelectedIndex=0;
        imgcbxColorRamp.SelectedIndex=0;
    }
}
```

(2)输入图层选择响应函数，步骤如下：

①先取第一波段，确定栅格数据是否存在属性表；

②如果不存在，程序试图重建属性表；

③创建分级数据组，如果属性表存在，使用 CreateBreakClass( )函数，不存在使用 CreateStretchBreakClass( )函数；

④在数据表上显示分级区间；

⑤激活色带选择事件。

代码如下：

```
private void ListBoxLayers_SelectedIndexChanged(object sender, EventArgs e)
{
    try
    {
        string layerName=this.ListBoxLayers.SelectedItem.ToString();
        IRasterLayer rasterLayer=this.GetRasterLayer(layerName);

        //取得栅格数据的波段数据
```

```
            IRasterBandCollection pBandCol = rasterLayer.Raster as IRas-
terBandCollection;
            m_bandRaster = pBandCol.Item(0);
            m_bandRaster.HasTable(out _isTableExisted);

            //如果属性表不存在,尝试重建属性表
            if (!_isTableExisted)
            {
                IRasterProps rasterProps = (IRasterProps)rasterLayer.Raster;
                if (rasterProps.PixelType != rstPixelType.PT_DOUBLE &&
                    rasterProps.PixelType != rstPixelType.PT_FLOAT)
                {
                    IDataset pDataset = rasterLayer as IDataset;
                    _isTableExisted = RebuildRasterAttribute(pDataset.Work-
space,pDataset.Name);
                }
            }

            //创建分级数据
            int numDesiredClasses = this.cbxClassNumber.SelectedIndex;
            double[] clsValues = (this._isTableExisted) ?
                                        CreateBreakClass(m_bandRaster,
numDesiredClasses):
                                        CreateStretchBreakClass(m_ban-
dRaster,numDesiredClasses);
            //在数据表上显示分级区间
            DisplayValues(clsValues);

            //激活色带事件
            imgcbxColorRamp_SelectedIndexChanged(sender,e);
            btnApply.Enabled = true;
        }
        catch (System.Exception ex)
        {

        }
    }
```

(3)遍历 DataGridView 中【颜色】列，将单元背景色设置为符号颜色，代码如下：

```
    private void imgcbxColorRamp_SelectedIndexChanged(object sender,
EventArgs e)
    {
        if (imgcbxColorRamp.SelectedIndex < 0)
            return;
        IEnumColors pColorEnumerater =
                            CreateAlgorithmicColorRamp(this.cbxClass-
Number.SelectedIndex+1);

        //将色带颜色值显示在 DataGridView 中
        pColorEnumerater.Reset();
        for (int i=0; i < this.dataGridView1.RowCount; i++)
        {
            IColor pNextUniqueColor=pColorEnumerater.Next();
            if (pNextUniqueColor!=null)
                this.dataGridView1[0,i].Style.BackColor=
                                        ColorTranslator. FromOle
(pNextUniqueColor.RGB);
            else
            {
                pColorEnumerater.Reset(); i--;
            }
        }
    }

    private void cbxClassNumber_SelectedIndexChanged(object sender,
EventArgs e)
    {
        ListBoxLayers_SelectedIndexChanged(sender,e);
    }

    private void btnOk_Click(object sender,EventArgs e)
    {
        if (btnApply.Enabled==true)
        {
            btnApply_Click(null,null);
        }
        this.Close();
```

```
}

private void btnApply_Click(object sender,EventArgs e)
{
    RasterBreakClassRenderer();
}
```

3. 核心函数实现

CreateBreakClass(…)实现过程是：通过 IBasicHistogram 接口对栅格数据波段的属性表进行分析，获取值-频数数组，然后使用 EqualInterval 类进行等级划分，得到分级数组。

代码如下：

```
private double [ ] CreateBreakClass ( IRasterBand pBand, int num-
DesiredClasses)
{
    //构造 Histogram
    ITableHistogram pTableHistogram=new ESRI.ArcGIS.Carto.BasicT-
ableHistogramClass();
    IBasicHistogram pHistogram=pTableHistogram as IBasicHistogram;
    pTableHistogram.Field="Value";
    pTableHistogram.Table=pBand.AttributeTable;

    //提取值和频数数据组
    object dataFrequency=new object();
    object dataValues=new object();
    pHistogram.GetHistogram(out dataValues,out dataFrequency);

    //等间距分级
    IClassifyGEN pClassify=new EqualInterval();
    pClassify. Classify ( dataValues, dataFrequency, ref numDesired-
Classes);

    return pClassify.ClassBreaks as double[];
}
```

CreateStretchBreakClass(…)实现过程是：直接使用波段的 IBasicHistogram 属性，获取 0~255 值-频数数组，然后依据栅格数据的最大最小值，通过拉伸(或压缩)构造出实际的值-频数数组，再使用 EqualInterval 类进行等级划分，得到分级数组。

代码如下：

```
private double[ ] CreateStretchBreakClass(IRasterBand pRsBand,int
```

```
numDesiredClasses)
    {
        pRsBand.ComputeStatsAndHist();

        //获取 Bnd 0~255 值的频数数据组
        IRasterHistogram pRasterHistogram=pRsBand.Histogram;
        double[] dblCounts=pRasterHistogram.Counts as double[];
        int ValueCount=dblCounts.Length;

        //获取 Band 统计数据
        IRasterStatistics pRasterStatistic=pRsBand.Statistics;
        double dMaxValue=pRasterStatistic.Maximum;
        double dMinValue=pRasterStatistic.Minimum;

        //构造值数组
        double[] dblValues=new double[ValueCount];
        {
            double BinInterval=Convert.ToDouble((dMaxValue - dMinValue)/
ValueCount);
            for (int i=0; i < ValueCount; i++)
            {
                dblValues[i]=i*BinInterval+pRasterStatistic.Minimum;
            }
        }

        IClassifyGEN pClassify=new GeometricalIntervalClass();
        pClassify.Classify(dblValues,dblCounts,ref numDesiredClasses);

        return pClassify.ClassBreaks as double[];
    }
```

RasterBreakClassRenderer(…)是分级渲染的调度函数：

①获取分级数组；

②创建渐变色带；

③创建分级渲染器 RasterClassifyColorRampRendererClass；

④然后为不同分级配置不同颜色的符号；

⑤渲染器绑定到图层。

代码如下：

```
private void RasterBreakClassRenderer()
```

```
    {
        //获取分级数据
        double[] Classes=(this.isTableExisted) ?
                              CreateBreakClass(m_bandRaster,numDesiredClasses):
                              Classes=CreateStretchBreakClass(m_bandRaster,
numDesiredClasses);
        int ClassesCount=Classes.GetUpperBound(0);

        //创建色带
        IEnumColors pEnumColors=CreateAlgorithmicColorRamp(ClassesCount+1);

        //建立渲染器
        IRasterClassifyColorRampRenderer pClassBreaksRenderer=null;
        pClassBreaksRenderer=new RasterClassifyColorRampRendererClass();
        {
            //设置渲染器"值"字段属性,分级数量,排序规则
            pClassBreaksRenderer.ClassField="Value";
            pClassBreaksRenderer.ClassCount=ClassesCount+1;
            pClassBreaksRenderer.SortClassesAscending=true;
            //为不同分级值配置不同颜色的符号
            for (int breakIndex=0; breakIndex <=ClassesCount; breakIndex++)
            {
                ISimpleFillSymbol pFillSymbol=new SimpleFillSymbol();
                pFillSymbol.Color=pEnumColors.Next();
                pFillSymbol.Style=esriSimpleFillStyle.esriSFSSolid;

                pClassBreaksRenderer.set_Symbol(breakIndex,pFillSymbol as
ISymbol);
                pClassBreaksRenderer. set_Break ( breakIndex, Classes [ brea-
kIndex]);
                pClassBreaksRenderer. set_Label ( breakIndex, Classes [ brea-
kIndex].ToString());
            }
        }

        //渲染器绑定到图层
        string layerName=this.ListBoxLayers.SelectedItem.ToString();
        IRasterLayer rasterLayer=this.GetRasterLayer(layerName);
```

```
        rasterLayer.Renderer=pClassBreaksRenderer as IRasterRenderer;
        this.Close();
    }
```

4. 辅助函数

代码如下：

```
    //重建栅格数据属性表
    private bool RebuildRasterAttribute(IWorkspace workspace, string
rstFileName)
    {
        IRasterDataset rasterDataset=
                                (workspace as IRasterWorkspace).Open-
RasterDataset(rstFileName);
        IRasterDatasetEdit2 rasterDatasetEdit=(IRasterDatasetEdit2)
rasterDataset;
        rasterDatasetEdit.BuildAttributeTable();

        //IRasterBandCollection rasterBandCollection=rasterDataset as
IRasterBandCollection;
        //rasterBandCollection.SaveAs(fileName, workspace as IWork-
space,"TIFF");
        return true;
    }

    private void DisplayValues(double[] clsValues)
    {
        dataGridView1.Rows.Clear();
        for (int i=0; i < clsValues.Length - 1; i++)
        {
            //将属性值添加到表格
            object vntUniqueValue=clsValues[i].ToString("F6")+" - " +
                                clsValues[i+1].ToString("F6");;
            {
                dataGridView1.Rows.Add();
                dataGridView1[1,i].Value=vntUniqueValue;
                dataGridView1[2,i].Value=vntUniqueValue;
            }
        }
    }
```

```
    private IEnumColors CreateAlgorithmicColorRamp(int ClassesCount)
    {
        if (imgcbxColorRamp.SelectedIndex < 0)
            return null;

        IColor fromColor=((ItemEx)imgcbxColorRamp.SelectedItem).From-
Color;
        IColor toColor=((ItemEx)imgcbxColorRamp.SelectedItem).ToColor;

        //建立从黄到红的渐变色
        IAlgorithmicColorRamp pRamp=new AlgorithmicColorRamp();
        pRamp.Algorithm=ESRI.ArcGIS.Display.esriColorRampAlgorithm.
esriHSVAlgorithm;
        pRamp.FromColor=fromColor;
        pRamp.ToColor=toColor;
        pRamp.Size=ClassesCount;

        bool ok=false;
        pRamp.CreateRamp(out ok);
        return pRamp.Colors;
    }

    private IEnumLayer GetRasterLayers()
    {
        UID uid=new UIDClass();
        //uid.Value="{40A9E885-5533-11d0-98BE-00805F7CED21}";//Featu-
reLayer
        uid.Value="{D02371C7-35F7-11D2-B1F2-00C04F8EDEFF}"; //Raster-
Layer
        IEnumLayer layers=m_mapControl.Map.get_Layers(uid,true);

        return layers;
    }

    private IRasterLayer GetRasterLayer(string layerName)
    {
        IEnumLayer layers=GetRasterLayers();
```

```
        layers.Reset();

        ILayer layer=null;
        while ((layer=layers.Next())!=null)
        {
            if (layer.Name==layerName)
                return layer as IRasterLayer;
        }
        return null;
    }
```

此外，还涉及 InitColorRamp()辅助函数，请参阅之前有关内容。

8.4 调用分级符号渲染窗体

在【Theme】Tab 页上添加【Raster Graduate】按钮，建立 Click 响应函数，代码如下：

```
private voidbtnRasterGraduate_Click(object sender,EventArgs e)
{
    RasterGraduateSymbolizeFrm frm=new RasterGraduateSymbolizeFrm
(m_mapControl);
    if (frm.ShowDialog()==DialogResult.OK)
    {
        //更新主 Map 控件和图层控件
        this._AxMapControl.ActiveView.Refresh();
        this._AxMapControl.Update();
    }
}
```

8.5 编译运行

按下 F5 键，编译运行程序。

第 9 章　基于属性查询

9.1　知识要点

GIS 空间数据查询分为：基于属性特征的查询，基于空间关系的查询，联合查询。

ArcEngine 中 QueryFilterClass 类是一个依据属性约束条件的查询过滤器，IQueryFilter 是该类实现的主要接口，通过对 IQueryFilter 的 WhereClause 属性可设置任意复杂度的 SQL 条件子句，满足第一类查询条件的过滤要求。

ArcEngine 执行查询的接口主要有：IFeatureSelection 的函数(FeatureLayerClass 实现)，IFeatureClass 的函数(FeatureClass 实现)，前者使用 SelectFeatures()方法得到选择集，后者使用 Search()得到查询游标。

9.2　功能描述

单击【Query】Ribbon 页的【Query By Attribute】按钮，可弹出“查询”对话框，如图 9-1 所示。点击【+】按钮，组合关系表达式相关控件的内容构造一个 Where 子句，添加到查询条件文本框。添加多个字句时，以选定的布尔操作符连接。

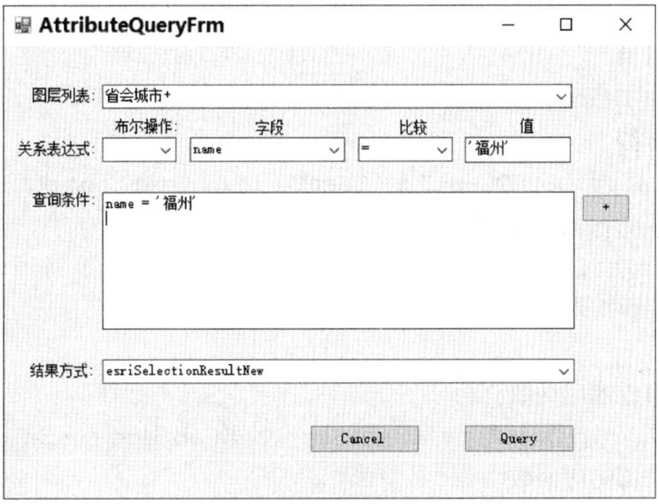

图 9-1　“查询”对话框

9.3 功能实现

1. 查询窗体设计

新建一个 Windows 窗体，命名为“QueryAttributeFrm. cs”，从工具箱拖动 5 个 ComBox（图层列表、查询方式等）、1 个 TextBox（填写查询条件）、2 个 Button（btnQuery、Cancel）控件到窗体。

表 9-1 **控件及其属性**

控　件	Name 属性	Text 属性	其　他
Combox	cbxLayers	选择图层	
Combox	cbxCompareOpr	比较操作符	设计时填充：=，>，>=，<，<=
Combox	cbxBoolOpr	布尔操作符	设计时填充：AND，OR
Combox	cbxFields	字段集	
Combox	cbxResultMethod	结果合成方式	
TextBox	txtWhereClause	Where 字句	
Button	btnWhereClause	添加 Where 字句	
Button	btnQuery	查询	
Button	btnCancel	取消	

代码如下：

```
public partial class QueryAttributeFrm:Form
{
    //私有成员
    private IMapControl3 m_mapControl=null;
    //构造函数
    public AttributeQueryFrm(IMapControl3 mapControl)
    {
        InitializeComponent();
        m_mapControl=mapControl;
    }
    //装载事件响应函数
    private void QuerAttributeyFrm_Load(object sender,EventArgs e)
    //查询按钮响应函数
    private void btnQuery_Click(object sender,EventArgs e)
}
```

2. 响应函数实现

(1)装载响应函数实现

窗体装载时完成两件事：一是用层名填充 cbxLayers，这里用到辅助函数 GetLayers()，它枚举器形式返回 Map 中所有矢量数据图层接口集合；二是用 Esri 选择结果枚举类型填充 cbxResultMethod。代码如下：

```
//装载事件响应函数
private void QueryAttributeFrm_Load(object sender,EventArgs e)
{
    //load all the feature layers in the map to the layers combo
    IEnumLayer layers=GetLayers();
    layers.Reset();
    ILayer layer=null;
    while ((layer=layers.Next())!=null)
    {
        cbxLayers.Items.Add(layer.Name);
    }
    //select the first layer
    if (cbxLayers.Items.Count > 0)
        cbxLayers.SelectedIndex=0;

    //RezultMethod
    this.cbxResultMethod.Items.Add(esriSelectionResultEnum.esriSe-
lectionResultNew.ToString());
    this.cbxResultMethod.Items.Add(esriSelectionResultEnum.esriS-
electionResultAdd.ToString());
    this.cbxResultMethod.Items.Add(esriSelectionResultEnum.esriS-
electionResultAnd.ToString());
    this.cbxResultMethod.Items.Add(esriSelectionResultEnum.esriS-
electionResultXOR.ToString());
    this.cbxResultMethod.Items.Add(esriSelectionResultEnum.esriS-
electionResultSubtract.ToString());
    cbxResultMethod.SelectedIndex=0;
}
```

(2)查询按钮 Click 响应函数

①根据层名利用辅助 GetFeatureLayer(...)函数，获取 IFeatureLayer 接口对象；

②转换要素选择接口：IFeatureSelection；

③创建 QueryFilter 查询过滤器，将 txtWhereClause 控件的内容(去掉换行符)赋值给

QueryFilter 的 WhereClase 属性；

④查询方式转换，将界面下拉框选择的结果合成方式字符串转换为枚举类型：esriSelectionResultEnum；此处用到辅助函数 ResultStringToEnum(...)；

⑤使用 IFeatureSelection 的 SelectFeatures()函数执行查询；

⑥为选择显示配置颜色；

⑦刷新选择集。

代码如下：

```
//执行查询
private void BtnQuery_Click(object sender,EventArgs e)
{
    IFeatureLayer pFeatureLayer = GetFeatureLayer((string)cbxLayers.
SelectedItem);
    if (null = =pFeatureLayer)
        return;

    IFeatureSelection pFeatureSelection =pFeatureLayer as IFeature-
Selection;
    ISelectionSet pSelectionSet =pFeatureSelection.SelectionSet;

    //创建 QueryFilter 空间过滤器对象
    IQueryFilter pQueryFilter =new QueryFilterClass();
    pQueryFilter.WhereClause = txtWhereClause.Text.Replace(" \r\
n"," ");

    //执行查询
    esriSelectionResultEnum resultMethod=StringToResultEnum((string)
cbxResultMethod.SelectedItem);
    pFeatureSelection.SelectFeatures(pQueryFilter, resultMethod,
false);

    ////为选择显示配置颜色
    //Color color=Color.FromArgb(255,0,0);
    //pFeatureSelection.SetSelectionSymbol=false;
    //pFeatureSelection.SelectionColor=ColorToIRgbColor(color);
    this.m_mapControl.ActiveView.PartialRefresh(esriViewDrawPhase.
esriViewGeoSelection,null,null);
}
```

(3)图层选择响应函数

此函数主要作用是将该图层中可用的字段填充到 Feild Combox。

```
private void cbxLayers_SelectedIndexChanged(object sender,EventArgs e)
{
    cbxFields.Items.Clear();
    IFeatureLayer pFeatureLayer = GetFeatureLayer(this.cbxLayers.
SelectedItem.ToString());
    if (pFeatureLayer!=null)//判断是否找到
    {
        IFields pFields=pFeatureLayer.FeatureClass.Fields;
        int nCount=pFields.FieldCount;
        //将该图层中可用的字段填充到 Feild Combox
        for (int i=0; i < nCount; i++)
        {
            IField pField=pFields.get_Field(i);
            if (pField.Type!=esriFieldType.esriFieldTypeGeometry &&
                pField.Type!=esriFieldType.esriFieldTypeBlob &&
                pField.Type!=esriFieldType.esriFieldTypeRaster)//判断
类型
            {

                cbxFields.Items.Add(pField.Name);
            }
        }
        cbxFields.SelectedIndex=0;
        cbxBoolOpr.SelectedIndex=0;
        cbxCompareOpr.SelectedIndex=0;
    }
}
```

(4)Where 子句添加到查询条件文本框

代码如下:

```
private void btnWhereClause_Click(object sender,EventArgs e)
{
    string strBoolOpr=this.cbxBoolOpr.SelectedItem.ToString() +" ";
    strBoolOpr=(strBoolOpr.IndexOf("  ")>=0) ? "":strBoolOpr;

    string strCompareOpr=this.cbxCompareOpr.SelectedItem.ToString()+" ";
    string strFieldName=this.cbxFields.SelectedItem.ToString()+" ";
```

```
        string strValue=this.txtValue.Text+" \r\n";
        string whereclause=strBoolOpr+strFieldName+strCompareOpr+str-
Value;

        this.txtWhereClause.Text +=whereclause;
    }
```

(5)辅助函数实现

①根据层类型 UID 获取矢量图层，函数代码如下：

```
    private IEnumLayer GetLayers()
    {
            UID uid=new UIDClass();
            uid.Value="{40A9E885-5533-11d0-98BE-00805F7CED21}";
            IEnumLayer layers=m_mapControl.Map.get_Layers(uid,true);
            return layers;
    }
```

②根据层名获取矢量图层,函数代码如下：

```
    private IFeatureLayerGetFeatureLayer(string layerName)
    {
            //get the layers from the maps
            IEnumLayer layers=GetLayers();
            layers.Reset();
            ILayer layer=null;
            while ((layer=layers.Next())!=null)
            {
                if (layer.Name==layerName)
                    return (layer as IFeatureLayer);
            }
            return null;
    }
```

③转换查询方式字符为相应枚举类型,代码如下：

```
    private esriSelectionResultEnumStringToResultEnum(string strMethod)
    {
        esriSelectionResultEnum result;
        switch( strMethod )
        {
            case "esriSelectionResultNew":
                result = esriSelectionResultEnum. esriSelectionResult-
```

```
New;
                break;
            case "esriSelectionResultAdd":
                result=esriSelectionResultEnum.esriSelectionResultAdd;
                break;
            case "esriSelectionResultSubtract":
                  result = esriSelectionResultEnum. esriSelectionResult-
Subtract;
                break;
            case "esriSelectionResultAnd":
                result=esriSelectionResultEnum.esriSelectionResultAnd;
                break;
            case "esriSelectionResultXOR":
                result=esriSelectionResultEnum.esriSelectionResultXOR;
                break;
            default:
                result=esriSelectionResultEnum.esriSelectionResultNew;
                break;
        }
        return result;
    }
```

9.4 调用查询窗体

在【Query】Tab 页添加【Query By Attribute】按钮，并建立 Click 响应函数如下；

```
private void btnQueryByAttrbute_Click(object sender,EventArgs e)
{
    QueryAttributeFrm queryfrm=new QueryAttributeFrm(m_mapContrl);
    Queryfrm.Show( );
}
```

9.5 编译运行

按下 F5 键，编译运行程序。

第10章　空间查询

10.1　知识要点

GIS空间数据查询分为：基于属性特征的查询，基于空间关系的查询，联合查询。

ArcEngine中SaptialFilterClass类是一个依据空间约束条件的查询过滤器，ISaptialFilter是该类实现的主要接口，通过对ISaptialFilter的Geometry属性可设置查询的参考空间对象，SpatialRel属性设置查询空间关系，满足第二类查询条件的过滤要求。

ISaptialFilter是通过继承IQueryFilter而来，因此：SaptialFilterClass对象同时具备IQueryFilter接口的功能，技术上SaptialFilterClass可满足上述第三类查询条件的过滤要求。但对于查询功能设计，ArcEngine联合查询是通过控制当前查询结果集与原有结果集(或称选择集)的合成方式(即：在当前的选择集中选择)实现。

使用ISaptialFilter实现空间关系查询的要点如下：

①由于ISaptialFilter只接受一个Geometry对象，因此使用ISaptialFilter实现空间关系查询时，若面对多个空间对象(例如一个选择集，一个要素类等)，就必须进行打包处理，使其成一个Geometry对象，可以使用IGeometryBag完成。

②确定待查询空间数据集与参考空间对象的空间关系，ArcEngine支持的空间关系包括Intersect，Within，Contain，Touch，Cross，Overlap等。

③确定查询结果集与原选择集之间的合成方式，ArcEngine支持的结果集合成方式包括：

a. 新建选择集（Select features in）；

b. 添加到当前的选择集（Adds to the current selection）；

c. 从当前的选择集中去除（Subtracts from the current selection）；

d. 在当前的选择集中选择（Selects from the current selection）。

10.2　功能描述

点击【Query】Tab页中【Query By Location】按钮，弹出如下空间查询对话框，如图10-1所示，具有类似于ArcMap的Select By Location的查询功能，即根据参考图层中的要素与目标图层的空间关系(如覆盖、相交等)，在目标图层中查询到符合要求的要素集，并高亮显示。

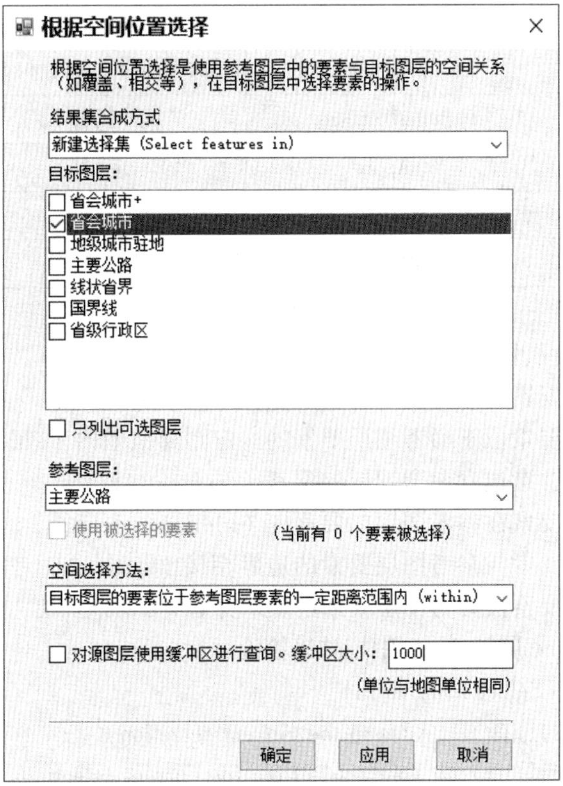

图 10-1　空间查询对话框

10.3　功能实现

10.3.1　类设计

1. 界面设计

新建一个 C# Windows 类，命名为“QueryBySpatialFrm. cs”。界面元素详见表 10-1。

表 10-1　**控件及其属性**

控　　件	Name 属性	Text 属性	其　他
Combox	cbxRefLayer	参考图层	
Combox	cbxSynthetizeMethod	结果集合成方式	
Combox	cbxSpatialRelation	空间关系	
CheckedList	checkedListBoxTargetLayers	目标图层	

续表

控　　件	Name 属性	Text 属性	其　他
CheckBox	checkBoxBuffer	参考图层应用缓冲区	
TextBox	txtBufferDistance	缓冲区大小	
Button	btnApply	应用	
Button	btnOK	确定	
Button	btnCancel	取消	

cbxSpatialRelation 属性 Items 中按顺序填充：

- 目标图层的要素与参考图层的要素相交（intersect）
- 目标图层的要素位于参考图层要素的一定距离范围内（within）
- 目标图层的要素包含参考图层的要素（contain）
- 目标图层的要素在参考图层的要素内（within）
- 目标图层的要素与参考图层要素的边界相接（touch）
- 目标图层的要素被参考图层要素的轮廓穿过（cross）

cbxSynthetizeMethod 属性 Items 中按顺序填充：

- 新建选择集（Select features in）
- 添加到当前的选择集（Adds to the current selection）
- 从当前的选择集中去除（Subtracts from the current selection）
- 在当前的选择集中选择（Selects from the current selection）

2. 类设计源代码

```
public class QueryBySpatialFrm
{
    IMapControl3 m_mapControl=null;
    IMap m_pMap=null;

    public QueryBySpatialFrm(IMapControl3 mapControl)
    {
        InitializeComponent();

        m_mapControl=mapControl;
        m_pMap=mapControl.Map;
    }

    //窗体加载时触发事件,执行函数
    private void QueryBySpatialFrm_Load(object sender,EventArgs e)
    //点击应用按钮时,执行函数
```

```
    private void btnApply_Click(object sender,EventArgs e)
    //点击确定按钮时,执行函数
    private void btnOK_Click(object sender,EventArgs e)

     //若干功能函数:
    ……

}
```

10.3.2 消息响应函数

1. btnApply_Click()函数实现方法

①将参考图层中所有要素几何体打包：这是实现本功能的关键点，用到功能函数 CreateGeometryUnion()。

②转换空间关系索引为相应的空间关系枚举：空间查询方法按照 ArcEngine 中枚举类型 esriSpatialRelEnum 填充在 cbxSpatialRelation 中。应用时，根据选择的索引值，使用辅助函数 IndexToSpatialRelation()转换成相应的枚举类型。

③创建查询过滤器：利用前两步的参数构造一个空间过滤器。

④将结果集合成方法索引转换为相应的结果集合成方法枚举：类似于②中情形，用到辅助 IndexToResultEnum()。

⑤执行选择查询函数；用功能函数 QueryByLocation()。

源代码如下：

```
private void btnApply_Click(object sender,EventArgs e)
{
    try
    {
        //获取参考图层中所有要素几何体的联合
        string strLayerName=cbxRefLayer.SelectedItem.ToString();
        IFeatureLayer pFeatureLayer=GetFeatureLayer(strLayerName);
        IGeometry pGeometry=CreateGeometryUnion(pFeatureLayer);

        //转换空间选择方法索引为相应的空间选择方法
        esriSpatialRelEnum enumSpatialRel=
                                    IndexToSpatialRelation(cbx-
SpatialRelation.SelectedIndex);

        //创建查询过滤器
        ISpatialFilter spatialFilter=new SpatialFilterClass();
        spatialFilter.Geometry=pGeometry;
        spatialFilter.SpatialRel=enumSpatialRel;
```

```
            //将结果集方法索引转换为相应的结果集方法
            esriSelectionResultEnum enumResultMethod =
                                            IndexToResultEnum ( cbxSynthe-
tizeMethod.SelectedIndex);

            //执行选择查询函数
            QueryByLocation(spatialFilter,enumResultMethod);
        }
        catch
        { }
    }
```

2. QueryBySpatialFrm_Load( )函数

窗体加载时触发事件，执行本函数。主要作用：

①用 Map 图层名填充 checkedListBoxTargetLayers，cbxRefLayer。

②设置 cbxRefLayer/ cbxSynthetizeMethod/ cbxSpatialRelation 初始值。

源代码如下：

```
    private void QueryBySpatialFrm_Load(object sender,EventArgs e)
    {
        try
        {
            //清空目标图层列表
            checkedListBoxTargetLayers.Items.Clear();
            string layerName;    //设置临时变量存储图层名称

            //对 Map 中的每个图层进行判断并添加图层名称
            for (int i = 0; i < m_pMap.LayerCount; i++)
            {
                //如果该图层为图层组类型,则分别对所包含的每个图层进行操作
                if (m_pMap.get_Layer(i) is GroupLayer)
                {
                    //使用 ICompositeLayer 接口进行遍历操作
                    ICompositeLayer compositeLayer = m_pMap.get_Layer(i)
as ICompositeLayer;
                    for (int j = 0; j < compositeLayer.Count; j++)
                    {
                        //将图层的名称添加到 checkedListBoxTargetLayers 控件和
cbxRefLayer 控件中
```

```
                    layerName=compositeLayer.get_Layer(j).Name;
                    checkedListBoxTargetLayers.Items.Add(layerName);
                    cbxRefLayer.Items.Add(layerName);
                }
            }
            //如果图层不是图层组类型,则直接添加名称
            else
            {
                layerName=m_pMap.get_Layer(i).Name;
                checkedListBoxTargetLayers.Items.Add(layerName);
                cbxRefLayer.Items.Add(layerName);
            }
        }

        //将 comboBoxSourceLayer 控件的默认选项设置为第一个图层的名称
        cbxRefLayer.SelectedIndex=0;
        //将 comboBoxMethods 控件的默认选项设置为第一种空间选择方法
        cbxSpatialRelation.SelectedIndex=0;
        //将 cboResultMethod 控件的默认选项设置为第一种空间选择方法
        cbxSynthetizeMethod.SelectedIndex=0;
    }
    catch { }
}
```

10.3.3　核心函数

1. CreateGeometryUnion()函数

由于查询过滤器只接受 IGeometry 参数，因此需要将参考图层所有要素的几何体合并为一个 IGeometry，CreateGeometryUnion()函数可实现这个功能。步骤如下：

①创建几何体包 GeometryBag 对象。

②然后遍历参考图层要素，使用 IGeometryCollection 接口将几何体逐个加到几何体包中，如果使用缓冲区则将要素缓冲区加到包中。

③初始化一个与参考图层几何类型同类型的几何对象，如果使用缓冲区则一定是 Polygon 类型，见辅助函数 InitializeGeometry(esriGeometryType)。

④将初始几何对象转换为 ITopologicalOperator 接口，并进行 ConstructUnion 操作。

源代码如下：

```
private IGeometry CreateGeometryUnion(IFeatureLayer featureLayer)
{
    //使用 null 作为查询过滤器得到图层中所有要素的游标
```

```
    IFeatureCursor featureCursor=featureLayer.Search(null,false);

    //建立几何体包,存储每一个源要素的几何体
    IGeometryCollection pGeoCols = new GeometryBag( ) as IGeometry-
Collection;
    IFeature feature=featureCursor.NextFeature();
    while (feature!=null) //当游标不为空时
    {
        if (checkBoxBuffer.Checked)
        {
            //当前要素的几何体缓冲区
            ITopologicalOperator topo=feature.Shape as ITopologicalO-
perator;
            double dBufferDistance=double.Parse(txtBufferDistance.Text);
            IGeometry pBuffer=topo.Buffer(dBufferDistance);
            pGeoCols.AddGeometry(pBuffer);
        }
        else
        {
            pGeoCols.AddGeometry(feature.Shape);
        }

        //移动游标到下一个要素
        feature=featureCursor.NextFeature();
    }

    //初始化 IGeometry 接口的对象
    IGeometry geometry = InitializeGeometry(featureLayer.Feature-
Class.ShapeType);
    //使用 ITopologicalOperator 接口进行几何体的拓扑操作
    ITopologicalOperator topologicalOperator=geometry as ITopolog-
icalOperator;

    //执行 ConstructUnion 操作,将几何体包与初始几何体合并
    topologicalOperator.ConstructUnion(pGeoCols as IEnumGeometry);

    //返回最新合并后的几何体
    return geometry;
```

```
}

//初始化 IGeometry 接口的对象
private IGeometryInitializeGeometry(esriGeometryType shapeType)
{
    if(checkBoxBuffer.Checked)
        return (new PolygonClass());

    switch (shapeType)
    {
        case esriGeometryType.esriGeometryPoint:
            return (new PointClass());
            break;
        case esriGeometryType.esriGeometryPolyline:
            return (new PolylineClass());
            break;
        case esriGeometryType.esriGeometryPolygon:
            return (new PolygonClass());
            break;
    }

    return null;
}
```

2. QueryByLocation()函数

遍历被选择的目标图层，并对每一个图层进行空间查询操作，代码如下：

```
/// <summary>
///根据已配置的查询条件来执行空间查询操作
/// </summary>
private void QueryByLocation(ISpatialFilter spatialFilter,esriSe-
lectionResultEnum resultMethod)
{
    IFeatureLayer featureLayer;
    //对所选择的目标图层进行遍历,并对每一个图层进行空间查询操作
    for (int i=0; i < checkedListBoxTargetLayers.CheckedItems.Count;
i++)
    {
        //根据选择的目标图层名称获得对应的矢量图层
        featureLayer = GetFeatureLayer((string) checkedListBoxTar-
```

```
getLayers.CheckedItems[i]);
            //获取 IFeatureSelection 接口
            IFeatureSelection featureSelection = featureLayer as IFeature-
Selection;

            //使用 IFeatureSelection 的 SelectFeatures 方法执行查询,将结果合成到选择
集中
            featureSelection.SelectFeatures((IQueryFilter)spatialFil-
ter,resultMethod,false);
        }

        //使用 IActiveView 接口进行视图刷新(仅对选择要素)
        IActiveView activeView=m_pMap as IActiveView;
        activeView.PartialRefresh(esriViewDrawPhase.esriViewGeoSelec-
tion,null,activeView.Extent);
    }
```

10.3.4 辅助函数

```
    /// <summary>
    ///转换空间关系索引为相应的空间关系枚举类型
    /// </summary>
    private esriSpatialRelEnum IndexToSpatialRelation(int iSpatialRe-
lIndex)
    {
        esriSpatialRelEnum pSpatialRel;
        switch (iSpatialRelIndex )
        {
            case 0://目标图层的要素与参考图层的要素相交 (intersect)
                pSpatialRel=esriSpatialRelEnum.esriSpatialRelIntersects;
                break;
            case 1://目标图层的要素位于参考图层要素的一定距离范围内 (within)
                pSpatialRel=esriSpatialRelEnum.esriSpatialRelContains;
                break;
            case 2://目标图层的要素包含参考图层的要素 (contain)
                pSpatialRel=esriSpatialRelEnum.esriSpatialRelWithin;
                break;
            case 3://目标图层的要素在参考图层的要素内 (within)
                pSpatialRel=esriSpatialRelEnum.esriSpatialRelContains;
```

```
            break;
        case 4://目标图层的要素与参考图层要素的边界相接（touch）
            pSpatialRel=esriSpatialRelEnum.esriSpatialRelTouches;
            break;
        case 5://目标图层的要素被参考图层要素的轮廓穿过（cross）
            pSpatialRel=esriSpatialRelEnum.esriSpatialRelCrosses;
            break;
        default:
            pSpatialRel=esriSpatialRelEnum.esriSpatialRelIntersects;
            break;
    }

    return pSpatialRel;
}
/// <summary>
///转换结果集合成方式索引为相应的结果集枚举类型
/// </summary>
private esriSelectionResultEnum IndexToResultEnum(int methodIndex)
{
    esriSelectionResultEnum result;
    switch (methodIndex)
    {
        case 0://新建选择集（Select features in）:
            result=esriSelectionResultEnum.esriSelectionResultNew;
            break;
        case 1://添加到当前的选择集（Adds to the current selection）:
            result=esriSelectionResultEnum.esriSelectionResultAdd;
            break;
        case 2://从当前的选择集中去除（Subtracts from the current selection）:
            result=esriSelectionResultEnum.esriSelectionResultSubtract;
            break;
        case 3://在当前的选择集中选择（Selects from the current selection）:
            result=esriSelectionResultEnum.esriSelectionResultAnd;
            break;
        //case "esriSelectionResultXOR":
        //    result=esriSelectionResultEnum.esriSelectionResultXOR;
        //    break;
        default:
```

```
            result = esriSelectionResultEnum.esriSelectionResultNew;
            break;
        }
        return result;
    }
```

10.4 功能调用

在【Query】Tab 页中，添加【Query By Location】按钮，建立 Click 响应函数；

```
private void btnQueryByLocation_Click(object sender,EventArgs e)
{
    QueryBySpatialFrm frm = new QueryBySpatialFrm(_mapControl);
    frm.Show();
}
```

10.5 编译测试

按下 F5 键，编译运行程序。

第 11 章　缓冲区分析(使用 GP 工具)

11.1　知识要点

缓冲区分析(Buffer)是对选中的一组或一类地图要素(点、线或面)按设定的距离条件，围绕其要素而形成一定距离的缓冲多边形实体，从而实现在二维空间研究实体的影响范围的分析方法。缓冲区应用的实例如：污染源对其周围的污染量随距离而减小，确定污染的区域；为失火建筑找到距其 500 米范围内所有的消防水管等。

从 ArcGIS 9.3 后，ArcEngine 提供 GP 工具分析方法，极大降低了用户的开发难度，使用 GP 工具的步骤是：

①构建一个 Geoprocessor 的类对象 GP，将由它来执行 Geoprocessing 的工具。

②构建一个 Geoprocessing 工具的类对象，比如这里的 Intersect 工具的类对象 Intersect。

③为工具填写参数，参数分 in 和 out，Required 和 Optional。

④调用 GP 的 Execute 方法执行 Geoprocessing 工具。

使用 GP 工具开发的难点是为工具配置参数，有关 GP 分析的更多知识请参看叠加分析。本章将介绍使用 GP 工具进行缓冲区分析的过程，还将介绍使用 BaseCommand 扩展功能进行命令包装的方法。

11.2　功能描述

点击【Spatial Analysis】Tab 页【Buffer】按钮，弹出如下缓冲区分析对话框，操作界面如图 11-1 所示。

操作界面包括缓冲图层选择、缓冲距离及单位、输出文件路径、分析过程信息显示等。

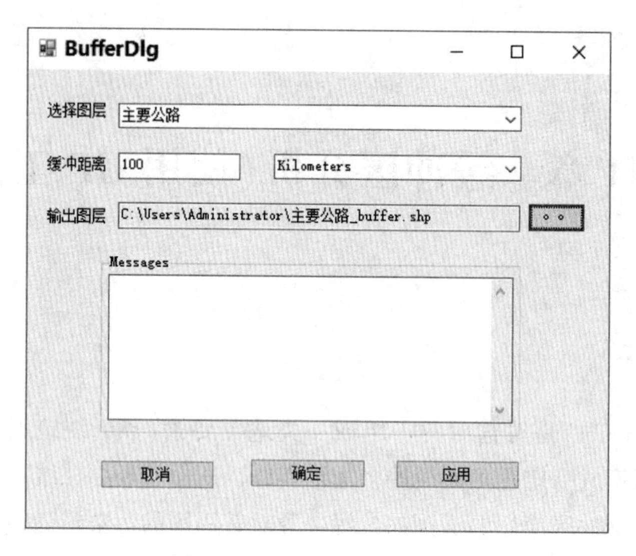

图 11-1 “Buffer Dlg”对话框

11.3 功能实现

11.3.1 新建功能窗体

1. 界面设计

项目中添加一个新的窗体，名称为“BufferAnalystFrm”，Name 属性设为“缓冲区分析”，添加 2 个 ComboBox、3 个 TextBox、4 个 Button 控件，设置控件的相关属性见表 11-1。

表 11-1 **控件及其属性、含义**

控 件	Name 属性	含 义	其 他
Combox	cbxLayers	选择图层	
Combox	cbxUnits	单位	
TextBox	txtBufferDistance	距离	
TextBox	txtOutputPath	输出路径	ReadOnly
TextBox	txtMessages	信息	MultiLine
Button	btnOutputLayer	输出路径设置	
Button	btnAnalyst	应用	
Button	btnCancel	关闭	
Button	btnOK	确定	DialogRezult. OK

2. 类结构设计

修改类定义代码：

```
public partial class BufferAnalystFrm:Form
{
    //添加类成员变量
    private IMap m_pMap=null;
    private const uint WM_VSCROLL=0x0115;
    private const uint SB_BOTTOM=7;
    //构造函数
    public BufferAnalystFrm(IMap pMap)
    {
        InitializeComponent();
        m_pMap=pMap;
    }
    //窗体加载时触发事件,执行函数
    private void BufferAnalystFrm_Load(object sender,EventArgs e)
    //点击输出路径按钮时,执行函数
    private void btnOutputLayer_Click(object sender,EventArgs e)
    //点击取消按钮时,执行函数
    private void btnCancel_Click(object sender,EventArgs e)
    //点击分析按钮时,执行函数
    private void btnAnalyst_Click(object sender,EventArgs e)
    //点击确定按钮时,执行函数
    private void btnOK_Click(object sender,EventArgs e)

    //若干功能函数:
    ……
}
```

11.3.2 消息响应函数

1. 载入响应函数 BufferAnalystFrm_Load()

BufferAnalystFrm 在载入时需要做几件事：

①用 Map 图层名填充 cbxLayers;

②加载 14 种 esri 单位填充 cbxUnits;

③设置缓冲区文件的默认输出路径，这里我们将默认输出路径设为系统临时目录。

代码如下：

```
private void BufferAnalystFrm_Load(object sender,EventArgs e)
{
```

```
        if (null==m_pMap || 0==m_pMap.LayerCount)
            return;

        //load all the feature layers in the map to the layers combo
        IEnumLayer layers=GetFeatureLayers();
        layers.Reset();
        ILayer layer=null;
        while ((layer=layers.Next())!=null)
        {
            cboLayers.Items.Add(layer.Name);
        }
        //select the first layer
        if (cboLayers.Items.Count > 0)
            cboLayers.SelectedIndex=0;

        string tempDir=System.IO.Path.GetTempPath();
        txtOutputPath.Text=System.IO.Path.Combine(tempDir,
                                        ((string)cboLayers.Selecte-
dItem+"_buffer.shp"));

        //set cboUnits
        cboUnits.Items.Add(esriUnits.esriCentimeters.ToString().Sub-
string(4));
        cboUnits.Items.Add(esriUnits.esriDecimalDegrees.ToString().
Substring(4));
        cboUnits.Items.Add(esriUnits.esriDecimeters.ToString().Sub-
string(4));
        cboUnits.Items.Add(esriUnits.esriFeet.ToString().Substring
(4));
        cboUnits.Items.Add(esriUnits.esriInches.ToString().Substring
(4));
        cboUnits.Items.Add(esriUnits.esriKilometers.ToString().Sub-
string(4));
        cboUnits.Items.Add(esriUnits.esriMeters.ToString().Substring
(4));
        cboUnits.Items.Add(esriUnits.esriMiles.ToString().Substring
(4));
        cboUnits.Items.Add(esriUnits.esriMillimeters.ToString().Sub-
```

```
string(4));
        cboUnits.Items.Add(esriUnits.esriNauticalMiles.ToString().
Substring(4));
        cboUnits.Items.Add(esriUnits.esriPoints.ToString().Substring
(4));
        cboUnits.Items.Add(esriUnits.esriUnitsLast.ToString().Sub-
string(4));
        cboUnits.Items.Add(esriUnits.esriUnknownUnits.ToString().
Substring(4));
        cboUnits.Items.Add(esriUnits.esriYards.ToString().Substring
(4));

        //set the default units of the buffer
        int units=Convert.ToInt32(m_pMap.MapUnits);
        cboUnits.SelectedIndex=units;
    }
```

2. 输出路径设置响应函数 btnOutputLayer_Click()

输出路径设置由 SaveFileDialog 实现，添加代码如下：

```
private void btnOutputLayer_Click(object sender,EventArgs e)
{
    //set the output layer
    SaveFileDialog saveDlg=new SaveFileDialog();
    saveDlg.CheckPathExists=true;
    saveDlg.Filter="Shapefile (*.shp)|*.shp";
    saveDlg.OverwritePrompt=true;
    saveDlg.Title="Output Layer";
    saveDlg.RestoreDirectory=true;
    saveDlg.FileName=(string)cboLayers.SelectedItem+"_buffer.shp";

    DialogResult dr=saveDlg.ShowDialog();
    if (dr==DialogResult.OK)
        txtOutputPath.Text=saveDlg.FileName;
}
```

3. 分析响应函数 btnAnalyst_Click()

分析按钮响应函数，激活缓冲区分析操作。步骤如下：

①有效性验证：如果不满足计算条件，就直接返回，由函数 ValidateCheck()实现；

②创建 Geoprocessor 代理类；

③创建 GPProcess 缓冲区分析工具；

④配置工具参数；

⑤运行 GPProcess 工具；

⑥显示叠置分析处理过程消息。

代码如下：

```
private void btnAnalyst_Click(object sender,EventArgs e)
{
    if (!ValidateCheck())
        return;
    //转换 txtBufferDistance 为 double 类型
    double bufferDistance;
    double.TryParse(txtBufferDistance.Text,out bufferDistance);
    //get the layer from the map
    IFeatureLayer layer=GetFeatureLayer((string)cboLayers.Select-
edItem);

    //scroll the textbox to the bottom
    ScrollToBottom();
    txtMessages.Text +=" \r\n 分析开始,这可能需要几分钟时间,请稍候.. \r\n";
    txtMessages.Update();

    //1:get an instance of the geoprocessor
    Geoprocessor gp=new Geoprocessor();
    gp.OverwriteOutput=true;

    //2:create a new instance of a buffer tool
    ESRI.ArcGIS.AnalysisTools.Buffer buffer=new Buffer();

    //3:set paramiter of tool
    buffer.in_features=layer;
    buffer.out_feature_class=txtOutputPath.Text;
    buffer.buffer_distance_or_field=Convert.ToString (bufferDis-
tance)+" "+(string)cboUnits.SelectedItem;
    buffer.dissolve_option="ALL";//这个要设成 ALL,否则相交部分不会
融合
    //buffer.line_side="FULL";默认是"FULL",最好不要改,否则出错
    //buffer.line_end_type="ROUND";默认是"ROUND",最好不要改,否则
出错
```

```
    //4:execute the buffer tool (very easy:-))
    IGeoProcessorResult results=null;
    try
    {
        results=(IGeoProcessorResult)gp.Execute(buffer,null);
    }
    catch (Exception ex)
    {
        txtMessages.Text +="Failed to buffer layer:"+layer.Name+" \r\n";
    }

    if (results.Status!=esriJobStatus.esriJobSucceeded)
    {
        txtMessages.Text +="Failed to buffer layer:"+layer.Name+" \r\n";
    }

    //scroll the textbox to the bottom
    ScrollToBottom();
    txtMessages.Text +=" \r\n 分析完成. \r\n";
    txtMessages.Text +="------------------------------------------------------- \r\n";
    //scroll the textbox to the bottom
    ScrollToBottom();

}
```

11.3.3 辅助函数

代码如下：

```
//根据层名获取要素层
private IFeatureLayer GetFeatureLayer(string layerName)
{
    //get the layers from the maps
    IEnumLayer layers=GetFeatureLayers();
    layers.Reset();

    ILayer layer=null;
    while ((layer=layers.Next())!=null)
    {
        if (layer.Name==layerName)
```

```
            return layer as IFeatureLayer;
        }

        return null;
    }
    //获取所有要素层的枚举器
    private IEnumLayer GetFeatureLayers()
    {
        UID uid=new UIDClass();
        uid.Value="{40A9E885-5533-11d0-98BE-00805F7CED21}";
        IEnumLayer layers=m_pMap.get_Layers(uid,true);

        return layers;
    }
```

11.4 功能调用

11.4.1 直接调用

在【Spatial Analysis】Tab 页添加【Buffer】按钮，建立 Click 响应函数，代码如下：

```
private void btnBuffer_Click(object sender,EventArgs e)
{
    BufferAnalystFrm frm=new BufferAnalystFrm (m_mapControl.Map);
    frm.Show();
}
```

11.4.2 包装成命令

如果 Buffer 功能包装成 ArcEngine 中 BaseTool 或 BaseCommand 的派生类，则可按命令方式调用，基类 BaseTool 和 BaseCommand 的区别是前者支持鼠标事件，在命令启动后可进行人机交互操作，例如在地图上拾取位置点等，后者不支持。Buffer 不需要人机交互，所以从 BaseCommand 继承。

1. 建立缓冲区命令类 BufferCommand

①新建 BufferCommand 底稿：选择添加新项，在 ArcGIS 目录下 Extending ArcObjects 中选择 Base Command 模板，如图 11-2 所示。

②修改构造函数 BufferCommand()：修改成员变量的值(base. m_category、base. m_caption 等)，使类库功能名称正确。

③重写 OnClick()函数：在其中启动 BufferAnalistFrm 对话框。

代码如下：

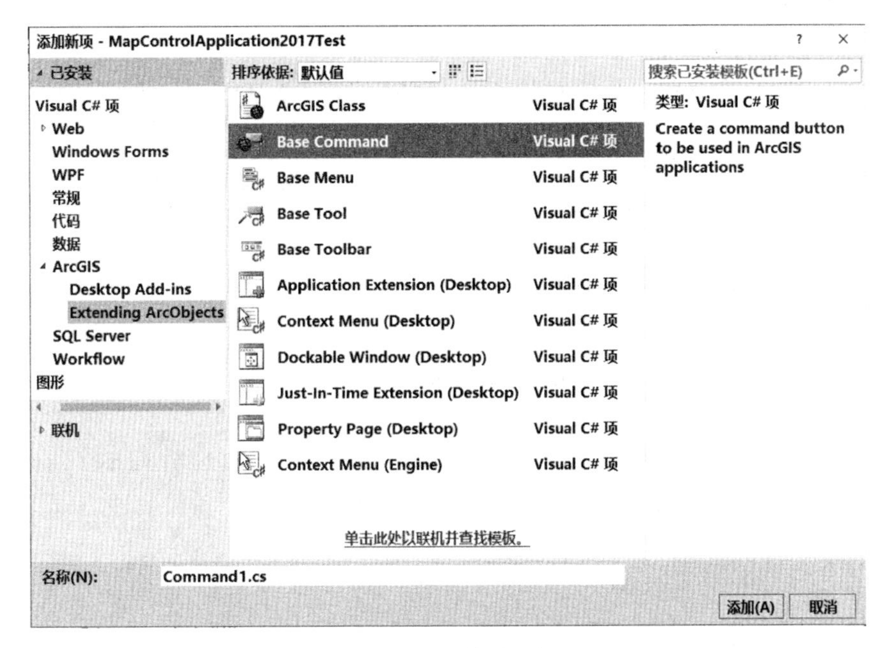

图 11-2　添加新项

```
public sealed classBufferCommand:BaseCommand
{
    #region COM Registration Function(s)
    //......
    #endregion

    private IHookHelper m_hookHelper=null;
    public BufferCmd()
    {
        base.m_category="自定义工具集";
        base.m_caption="缓冲区分析";
        base.m_message="缓冲区分析";
        base.m_toolTip="缓冲区分析";
        base.m_name="GpBufferLayer_BufferSelectedLayerCmd";

        try
        {
            string bitmapResourceName=GetType().Name+".bmp";
            base.m_bitmap=new Bitmap(GetType(),bitmapResourceName);
        }
```

```
            catch (Exception ex)
            {
                System.Diagnostics.Trace.WriteLine(ex.Message," Invalid
Bitmap");
            }
        }

        #region Overriden Class Methods
        /// <summary>
        /// Occurs when this command is created
        /// </summary>
        /// <param name="hook">Instance of the application</param>
        public override void OnCreate(object hook)
        {
            if (hook==null)
                return;

            if (m_hookHelper==null)
                m_hookHelper=new HookHelperClass();

            m_hookHelper.Hook=hook;
        }
        /// <summary>
        /// Occurs when this command is clicked
        /// </summary>
        public override void OnClick()
        {
            if (null==m_hookHelper)
                return;
            if (m_hookHelper.FocusMap.LayerCount > 0)
            {
                BufferAnalistFrm bufferDlg=new BufferAnalistFrm(m_hookHelper.
FocusMap);
                bufferDlg.Show();
            }
        }
        #endregion
    }
```

2. 调用自定义命令

代码如下：

```
private void btnBuffer_Click(object sender,EventArgs e)
{
    BufferCommand pCommand=new BufferCommand();
    pCommand.OnCreate(_mapControl.Object);
    pCommand.OnClick();
}
```

11.5　程序测试

①启动“缓冲区分析按钮”，弹出缓冲区分析对话框。

②选择缓存分析的图层，选择距离及单位，设置输出的图层。

③点击分析按钮，当出现“分析完成”字样时，工作完成。

第 12 章　矢量数据叠置分析

12.1　知识要点

叠置分析是 GIS 中一种常见的分析功能，它是将有关主题的各个数据层面进行叠置产生一个新的数据的分析方法，其结果综合了原来两个或多个层面要素所具有的属性，同时叠置分析不仅生成了新的空间关系，而且还将输入的多个数据层的属性联系起来产生了新的属性关系。

叠置分析是拓扑运算的范畴。拓扑运算包括：缓冲区(Buffer)、裁切(Clip)、凸多边形(ConvexHull)、切割(Cut)、差(Difference)、交集(Intersect)、对称差分(又称为异或，SymmetricDifference)和并集(Union)等，这些拓扑运算在 ArcEngine 的 ITopologicalOperator 接口中定义，高级几何对象(构成要素的几何对象：Multipoint、Point，Polygon 和 Polyline 等)都实现了这个接口；如果要使用在低等级的几何对象上(如 Segment、Path 或 Ring)，需要将它们组合成高级别几何对象。

ITopologicalOperator 接口是面向单个要素的。ArcEngine 还提供了面向数据集的接口：IBasicGeoprocessor 提供了 Clip，Dissolve，Intersect，Union，Merge 等方法。为降低 ArcObjects 开发者的复杂度。从 ArcGIS 9.3 后，ArcEngine 提供 GP 工具分析方法：其中 Geoprocessing 组件提供了数据分析、数据管理和数据转换等上百个 geoprocessing 工具；由 Geoprocessor对象可以方便地调用 Geoprocessing 中提供的各类工具。它是执行 ArcGIS 中 Geoprocessing 工具的唯一访问点。

使用 GP 工具的步骤是：

①构建一个 Geoprocessor 的类对象 GP，将由它来执行 Geoprocessing 的工具。

②构建一个 Geoprocessing 工具的类对象，比如这里的 Intersect 工具的类对象 Intersect。

③为工具填写参数，参数分 in 和 out，Required 和 Optional。

④调用 GP 的 Execute 方法执行 Geoprocessing 工具。

本章介绍使用 GP 工具进行叠加分析的过程。

12.2　功能描述

在【Spatial Analysis】Tab 页中单击【Overlay】按钮，弹出如下叠加分析对话框：即根据输入图层的要素与叠加图层的要素，进行 Union，Intersect，Identify，Erase 等操作，生成新的要素集。操作界面如图 12-1 所示。

图 12-1　Overlay Analysis Frm 操作界面

12.3　功能实现

12.3.1　新建功能窗体

1. 界面设计

项目中添加一个新的窗体，名称为"OverlayAnalysisFrm"，Name 属性设为"叠置分析"，添加 3 个 ComboBox、2 个 TextBox、4 个 Button 控件和一个 GroupBox，控件属性设置见表 12-1。

表 12-1　　**控件及其属性、含义**

控件类型	Name 属性	含　义	备　注
ComBox	cbxInputLayers	输入要素	
ComBox	cbxOverlayLayers	叠置要素	
ComBox	cbxOverLayMethod	叠置分析的方式	
TextBox	txtOutputPath	叠置结果的输出路径	
TextBox	txtMessage	叠置分析处理过程消息	Multiline 属性设为 True，ScrollBars 属性设为 Vertical，Dock 属性设为 Fill

续表

控件类型	Name 属性	含　义	备　注
Button	btnOutputLayer	选择输出路径	
Button	btnAnalist	进行叠置分析	
Button	btnCancel	取消	
Button	btnOK	确定	
GroupBox			作为 txtMessage 的容器

2. 类结构设计

修改类定义代码：

```
public class OverlayAnalysisFrm
{
    private IMap m_pMap=null;
    publicOverlayAnalysisFrm(IMap pMap)
    {
        InitializeComponent();
        m_pMap=pMap;
    }

    //窗体加载时触发事件,执行函数
    private voidOverlayAnalysisFrm_Load(object sender,EventArgs e)
    //点击应用按钮时,执行函数
    private void btnAnalyst_Click(object sender,EventArgs e)
    //点击确定按钮时,执行函数
    private void btnOK_Click(object sender,EventArgs e)
    //点击输出路径按钮时,执行函数
    private void btnOutputLayer_Click(object sender,EventArgs e)

    //若干功能函数
    private void resultOverlay()
    ……
}
```

12.3.2　消息响应函数

1. 载入响应函数 OverlayAnalysisFrm_Load()

OverlayAnalysisFrm 在载入时需要做几件事：

①用 Map 图层名填充 cbxInputLayers，cbxOverlayLayers；

②加载 5 种叠置方式到 cbxOverlayMethod 中；

③设置缓冲区文件的默认输出路径，这里我们将默认输出路径设为系统临时目录。

代码如下：

```
private void OverlayAnalysisFrm_Load(object sender,EventArgs e)
{
    //load all the feature layers in the map to the layers combo
    IEnumLayer layers=GetFeatureLayers();
    layers.Reset();
    ILayer layer=null;
    while ((layer=layers.Next())!=null)
    {
        cbxInputLayers.Items.Add(layer.Name);
    }
    layers.Reset();
    while ((layer=layers.Next())!=null)
    {
        cbxOverlayLayers.Items.Add(layer.Name);
    }

    this.cbxOverlayMethod.Items.Add("求交(Intersect)");
    this.cbxOverlayMethod.Items.Add("裁剪(Clip)");
    this.cbxOverlayMethod.Items.Add("擦除(Erase)");
    this.cbxOverlayMethod.Items.Add("求并(Union)");
    this.cbxOverlayMethod.Items.Add("标识(Identity)");
    this.cbxOverlayMethod.SelectedIndex=0;

    //select the first layer
    if (cbxInputLayers.Items.Count > 0)
        cbxInputLayers.SelectedIndex=0;
    if (cbxOverlayLayers.Items.Count > 0)
        cbxOverlayLayers.SelectedIndex=0;

    string tempDir=System.IO.Path.GetTempPath();
    string strFileName=(string)cbxInputLayers.SelectedItem +
                          (string) cbxOverlayMethod.SelectedItem +"_
overlay.shp";
    txtOutputPath.Text=System.IO.Path.Combine(tempDir,strFileName);
}
```

2. 输出路径设置响应函数 btnOutputLayer_Click( )

输出路径设置由 SaveFileDialog 实现，添加代码如下：

```
private void btnOutputLayer_Click(object sender,EventArgs e)
{
    //set the output layer
    SaveFileDialog saveDlg=new SaveFileDialog();
    saveDlg.CheckPathExists=true;
    saveDlg.Filter="Shapefile (*.shp) |*.shp";
    saveDlg.OverwritePrompt=true;
    saveDlg.Title="Output Layer";
    saveDlg.RestoreDirectory=true;
    saveDlg.FileName=(string)cbxOverlayLayers.SelectedItem +
                    (string)this.cbxOverlayMethod.SelectedItem+"_
overlay.shp";

    DialogResult dr=saveDlg.ShowDialog();
    if (dr==DialogResult.OK)
        txtOutputPath.Text=saveDlg.FileName;
}
```

3. 分析响应函数 btnAnalyst_Click( )

分析响应函数负责执行指定类型的叠加分析操作。步骤如下：

①有效性验证：如果不满足计算条件，则直接返回，由函数 ValidateCheck( )实现；

②创建 Geoprocessor 代理类；

③创建 GPProcess 叠加分析工具，并配置好参数。由函数 CreateGeoprocessorTool( )实现；

④运行 GPProcess 工具；由辅助函数 RunTool( )实现；

⑤显示叠置分析处理过程消息。

代码如下：

```
private void btnAnalyst_Click(object sender,EventArgs e)
{
    ScrollToBottom(this.txtMassages);
    txtMassages.Text +=" \r\n 分析开始,这可能需要几分钟时间,请稍候.. \r\n";
    txtMassages.Update();
    //1:有效性验证
    if (!ValidateCheck())
        return;
    //2:Setting up the Geoprocessor
    Geoprocessor GP=new Geoprocessor();
```

```
        GP.OverwriteOutput=true;

        //3:create a new instance of a tool
        string strOverlay=cbxOverlayMethod.SelectedItem.ToString();
//叠加方式字符串
        IGPProcess gpTool=CreateGeoprocessorTool( strOverlay );

        //4:  runtool
        txtMassages.Text=RunTool(GP,gpTool,null);

        //5:scroll the textbox to the bottom
        ScrollToBottom(this.txtMassages);
        txtMassages.Text +=" \r\n 分析完成. \r\n";
        txtMassages.Text +="-------------------------------------------------- \r\n";
        //scroll the textbox to the bottom
        ScrollToBottom(this.txtMassages);
    }
```

12.3.3 核心函数

1. CreateGeoprocessorTool()函数

本函数根据叠加方式创建相应的 gp 工具。每个工具对应一个创建子函数，分别是：CreateIntersectTool()，CreateUnionTool()，CreateIdentityTool()，CreateClipTool()，CreateEraseTool()。

因为在 ArcGIS 的叠置分析中 Union 和 Intersect 两种工具可以针对两个以上的图层进行叠置运算，它们输入参数要打包成 IGpValueTableObject 接口，然后赋值给 in_features。Union 和 Intersect 参数打包方式是一致的。这里 ContructMultiParameter()函数负责打包。

而 Identity、Clip、Erase 工具是针对两个要素的运算，其实质是使用叠置要素对输入要素进行更新的一个过程。它们是针对输入要素 in_ features 和叠加要素(如 identity_features)分别赋值。

代码如下：

```
    private IGPProcess CreateGeoprocessorTool(string strOverlay)
    {
        IFeatureLayer inputLayer=GetFeatureLayer((string)cbxInputLay-
ers.SelectedItem);
        IFeatureLayer overlayLayer=GetFeatureLayer((string)cbxOverlay-
Layers.SelectedItem);
        string strOutputPath=this.txtOutputPath.Text.ToString();//输
出文件名
```

```
        //根据叠加方式创建 gp 工具
        IGpValueTableObject valTbl=null;
        IGPProcess gpTool=null;
        switch (strOverlay)
        {
            case "求交(Intersect)":
                valTbl=ContructMultiParameter(inputLayer,overlayLayer);
                gpTool=CreateIntersectTool(valTbl,strOutputPath);
                break;
            case "求并(Union)":
                valTbl=ContructMultiParameter(inputLayer,overlayLayer);
                gpTool=CreateUnionTool(valTbl,strOutputPath);
                break;
            case "标识(Identity)":
                gpTool=CreateIdentityTool(inputLayer,overlayLayer,strOut-
putPath);
                break;
            case "裁剪(Clip)":
                gpTool=CreateClipTool(inputLayer,overlayLayer,strOut-
putPath);
                break;
            case "擦除(Erase)":
                gpTool=CreateEraseTool(inputLayer,overlayLayer,strOut-
putPath);
                break;
            default:
                break;;
        }
        return gpTool;
    }

    private IGpValueTableObject ContructMultiParameter(IFeatureLayer
inputLayer,IFeatureLayer overlayLayer)
    {
        //准备多重输入参数
        IGpValueTableObject valTbl=new GpValueTableObjectClass();
        valTbl.SetColumns(2);
```

```
        object row="";
        row=inputLayer;
        valTbl.AddRow(ref row);
        row=overlayLayer;
        valTbl.AddRow(ref row);

        return valTbl;
    }

    private IGPProcess CreateIntersectTool (IGpValueTableObject valTbl,
                                            string strOutputPath)
    {
        //创建工具
        Intersect intersectTool=new Intersect();
        //设置工具参数
        intersectTool.in_features=valTbl; //设置输入要素
        intersectTool.join_attributes="ALL";
        intersectTool.output_type="INPUT";
        intersectTool.out_feature_class=strOutputPath;//设置输出路径
        //用文件列表作参数
        //intersect.in_features=@"F:/foshan/Data/wuqutu_b.shp;F:/
foshan/Data/world30.shp";
        //intersect.out_feature_class=@"E:/intersect.shp";
        //intersect.join_attributes="ONLY_FID";
        //返回
        return (intersectTool as IGPProcess);
    }

    private IGPProcess CreateUnionTool (IGpValueTableObject valTbl,
string strOutputPath)
    {
        //创建工具
        Union unionTool=new Union();
        //设置工具参数
        unionTool.in_features=valTbl;//设置输入要素
        unionTool.join_attributes="ALL";
        unionTool.out_feature_class="INPUT";
        unionTool.out_feature_class=strOutputPath;
```

```
        //返回
        return (unionTool as IGPProcess);
    }

    private IGPProcess CreateIdentityTool(IFeatureLayer inputLayer,IF-
eatureLayer overlayLayer,string strOutputPath)
    {
        //创建工具
        Identity identityTool=new Identity();
        //设置工具参数
        identityTool.in_features=inputLayer; //设置输入要素
        identityTool.identity_features=overlayLayer; //设置叠加要素
        identityTool.out_feature_class=strOutputPath;//设置输出路径
        //返回
        return (identityTool as IGPProcess);
    }

    private IGPProcess CreateClipTool(IFeatureLayer inputLayer,IFeatu-
reLayer overlayLayer,string strOutputPath)
    {
        //创建工具
        Erase clipTool=new Erase();
        //设置工具参数
        clipTool.in_features=inputLayer;
        clipTool.erase_features=overlayLayer;
        clipTool.out_feature_class=strOutputPath;
        //eraseLayer.cluster_tolerance="0.01";
        //返回
        return (clipTool as IGPProcess);
    }

    private IGPProcess CreateEraseTool(IFeatureLayer inputLayer,IFea-
tureLayer overlayLayer,string strOutputPath)
    {
        //创建工具
        Erase eraseTool=new Erase();
        //设置工具参数
        eraseTool.in_features=inputLayer;
```

```
        eraseTool.erase_features=overlayLayer;
        eraseTool.out_feature_class=strOutputPath;
        //eraseLayer.cluster_tolerance="0.01";
        //返回
        return (eraseTool as IGPProcess);
    }
```

12.3.4　辅助函数

代码如下：

```
    //运行 GP 工具
    private string RunTool(Geoprocessor geoprocessor,IGPProcess process,
ITrackCancel TC)
    {
        // Set the overwrite output option to true
        geoprocessor.OverwriteOutput=true;
        IGeoProcessorResult results=null;
        try
        {
            results=(IGeoProcessorResult)geoprocessor.Execute(process,
null);
            return ReturnMessages(geoprocessor);

        }
        catch (Exception err)
        {
            Console.WriteLine(err.Message);
            return ReturnMessages(geoprocessor);
        }
    }

    // Function for returning the tool messages.
    private string ReturnMessages(Geoprocessor gp)
    {
        StringBuilder sb=new StringBuilder();
        if (gp.MessageCount > 0)
        {
            for (int Count=0; Count <=gp.MessageCount -1; Count++)
            {
```

```
                System.Diagnostics.Trace.WriteLine(gp.GetMessage(Count));
                sb.AppendFormat("{0}\n",gp.GetMessage(Count));
            }
        }
        return sb.ToString();
    }
    //滚动消息窗口
    private void ScrollToBottom(TextBox textbox)
    {
        //PostMessage((IntPtr)textbox.Handle,WM_VSCROLL,(IntPtr)SB_
BOTTOM,(IntPtr)IntPtr.Zero);
    }
```

12.4 功能调用

在【Spatial Analysis】Tab 页添加【Overlay】按钮。建立 Click 响应函数;

```
    private void btnOverlay_Click(object sender,EventArgs e)
    {
        OverlayAnalysisFrm frm=new OverlayAnalysisFrm (_mapControl.Map);
        frm.Show();
    }
```

12.5 编译测试

按下 F5 键，编译运行程序，点击“Overlay”，弹出叠置分析参数设置窗口，指定分析要素层，并设置输出路径。

第 13 章　栅格数据重分类

13.1　知识要点

所谓重分类就是依据一定规则对原有栅格像元值重新分类，并按新分类赋予一组新值并输出。

在 ArcGIS Engine 中，RasterReclassOpClass 类实现了栅格数据的重分类。该类实现了两个主要的接口，分别是 IReclassOp，IRasterAnalysisEnvironment 接口。IReclassOp 接口包含了以下几种分类分析方法：

①Reclass：使用表重分类；

②ReclassByASCIIFile：使用 ASCII 文件重分类；

③Slice：分割；

④Lookup：查找表；

⑤ReclassByRemap：构造映射表进行重分类。

下面以构造映射表进行重分类为例介绍重分类的方法，其他分类方法请读者参阅 ArcGIS Engine 的帮助文档。

13.2　功能描述

单击【Spatial Analysis】Tab 页【Reclassify】按钮，弹出如下重分类对话框：即根据输入栅格数据图层，进行重分类操作，生成新的栅格数据集，操作界面如图 13-1 所示。

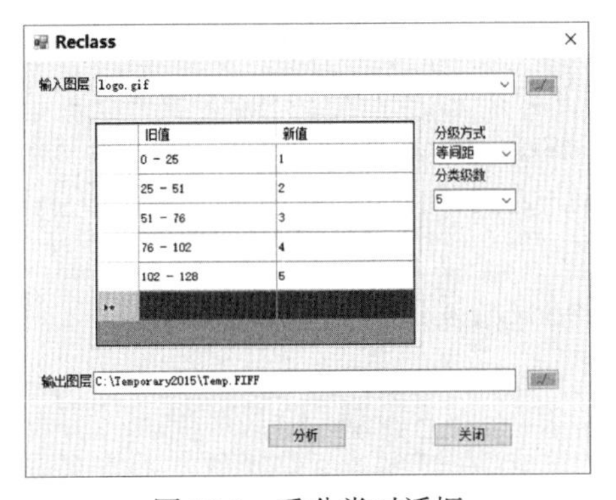

图 13-1　重分类对话框

13.3 功能实现

13.3.1 新建功能窗体

1. 界面设计

项目中添加一个新的窗体，名称为"ReclassifyFrm"，Name 属性设为"重分类"，添加 3 个 ComboBox、2 个 TextBox、2 个 Button 控件，1 个 DataGridView。

控件属性设置见表 13-1。

表 13-1　　控件及其属性、说明

控件类型	Name 属性	控件说明	备　注
ComBox	cbxInLayers	输入栅格数据	
ComBox	cbxCount	分类级数	
ComBox	cbxTypeCls	分类方式	
TextBox	txtInterval	间隔	
TextBox	txtOutLayers	输出结果文件名	
Button	btnOK	进行叠置分析	
Button	btnCancel	取消	
DataGridView	m_ pDataTable	展示映射表	

2. 类结构设计

添加如下代码，修改类定义：

```
public class ReclassifyFrm
{
    private IMapControl3 m_mapControl;
    private DataTable m_pDataTable=null;
    private ITable m_iArttributeTable=null;
    private IRasterStatistics m_rstStatistic=null;
    public ReclassifyFrm(IMapControl3 mapControl)
    {
        InitializeComponent();
        m_mapControl=mapControl;
        m_pDataTable=new DataTable();
    }
```

```
        //窗体加载时触发事件,执行函数
        private void ReclassifyFrm_Load(object sender,EventArgs e)
        //输入图层选择响应函数
        private void cbxInLayer_SelectedIndexChanged(object sender,
EventArgs e)
        //分类方式选择响应函数
        private void cbxTypeCls_SelectedIndexChanged(object sender,
EventArgs e)

        //点击响应函数,执行重分类操作
        private void btnAnalyst_Click(object sender,EventArgs e)
        //点击输出路径按钮时,执行函数
        private void btnOutLayer_Click(object sender,EventArgs e)
        //取消
        private void btnCancel_Click(object sender,EventArgs e)

        //若干功能函数
        private IRemap()
        private double[] CreateBreakClass(IRasterBand pBand,int num-
DesiredClasses)
        ……
    }
```

13.3.2 消息响应函数

1. 载入响应函数 ReclassifyFrm_Load()

ReclassifyFrm 在载入时需要做以下几件事：

①用 IMapControl3. Map 包含的图层名称填充 cbxInLayers；

②将六种分类方式名称(等间距、分位数、标准差、自然分割、几何间隔、自定义)填充到 cbxTypeCls 中，也可在设计时填充；

③设置输出文件的默认输出路径，这里我们将默认输出路径设为系统临时目录。

代码如下：

```
private void ReclassifyFrm_ Load(object sender, EventArgs e)
{
    IEnumLayer layers=GetRasterLayers();
    layers.Reset();
    ILayer layer=null;
    while ((layer=layers.Next())!=null)
    {
```

```
            cbxInLayer.Items.Add(layer.Name);
        }

        //设置 DataGridView 列信息
        DataColumn OldValue=new DataColumn("旧值", System.Type.GetType
("System.String"));
        DataColumn NewValue=new DataColumn("新值", System.Type.GetType
("System.Int32"));
        m_ pDataTable.Columns.Add(OldValue);
        m_ pDataTable.Columns.Add(NewValue);

        //其他初始化
        this.cbxCount.Text="5";
        this.txtOutLayer.Text=@"C: \Temporary2015\Temp.FIFF";

        this.cbxTypeCls.SelectedIndex=0;
        this.lblInterval.Visible=false;
        this.txtInterval.Visible=false;
    }
```

2. 输入图层选择响应函数

本函数被选择到某图层时必须为 m_iArttributeTable 赋值(为 m_rstStatistic 赋值均方差分级用得到)，以后计算以此为条件。

①先取第一波段，确定栅格数据是否存在属性表;

②如果不存在，程序试图重建属性表;

③如果重建失败(当数据类型为连续性数据类型)，则使用 getAttrTableByErgodic( )函数独立计算一个属性表。

具体代码如下:

```
    private void cbxInLayer_SelectedIndexChanged(object sender, Even-
tArgs e)
    {
        IRasterLayer rasterLayer=GetRasterLayer(cbxInLayer.Text);
        if (rasterLayer!=null)
        {
            IRasterProps rasterProps=(IRasterProps)rasterLayer.Raster;
            //取第一波段
            IRasterBandCollection pBandCol=rasterLayer.Raster as IRas-
terBandCollection;
            IRasterBand pBand=pBandCol.Item(0);
```

```
            //计算属性表
            bool bAttrTableExist;
            pBand.HasTable(out bAttrTableExist);
            if (bAttrTableExist)
            {
                m_iArttributeTable=pBand.AttributeTable;
            }
            else if (rasterProps.PixelType!=rstPixelType.PT_DOUBLE &&
                    rasterProps.PixelType!=rstPixelType.PT_FLOAT)
            {
                IDataset pDataset=rasterLayer as IDataset;
                RebuildRasterAttribute(pDataset.Workspace,pDataset.Name);
                m_iArttributeTable=pBand.AttributeTable;
            }
            else
            {
                //获取栅格属性表
                m_iArttributeTable=getAttrTableByErgodic(rasterLayer.Raster);
            }

            //获取 Raster 的 Statistics 信息
            pBand.ComputeStatsAndHist();
            IRasterHistogram pHistogram=pBand.Histogram;
            m_rstStatistic=pBand.Statistics;
            //IStatisticsResults statisticsRes=pHistogram as IStatistics-
Results;

            RemapRefresh(rasterProps.PixelType);
        }
    }
```

3. 输出路径设置响应函数 btnOutLayer_ Click()

输出路径设置由 SaveFileDialog 实现，添加代码如下：

```
    private void btnOutLayer_Click(object sender,EventArgs e)
    {
        SaveFileDialog saveDlg=new SaveFileDialog();
        saveDlg.CheckPathExists=true;
        saveDlg.Filter="Rasterfile (*.img) |*.img";
```

```
        saveDlg.OverwritePrompt=true;
        saveDlg.Title="Output Layer";
        saveDlg.RestoreDirectory=true;
        saveDlg.FileName=(string)cbxInLayer.SelectedItem+"_Reclass.img";

        DialogResult dr=saveDlg.ShowDialog();
        if (dr==DialogResult.OK)
            txtOutLayer.Text=saveDlg.FileName;
    }
```

4. 分析响应函数 btnAnalyst_Click()

分析响应函数负责执行指定分类方式的分类操作。步骤如下：

①根据 DataGridView 中定义的新旧数值对照关系，构造新旧值映射图(Remap)；

②创建重分类接口对象(IReclassOp)；

③执行重分类计算；

④结果保存到指定位置。

代码如下：

```
    private void btnAnalyst_Click(object sender,EventArgs e)
    {
        string layerName=this.cbxInLayer.SelectedItem.ToString();
        IRasterLayer rasterLayer=GetRasterLayer(layerName) as IRaster-
Layer;
        string outputFileName=this.txtOutLayer.Text;

        //构造 Remap
        IRemap pRemap=ConstructRemap();

        //创建 IReclassOp
        IReclassOp reclassOp;
        reclassOp=new RasterReclassOp() as IReclassOp;

        //执行重分类计算
        IGeoDataset rasDataset=rasterLayer.Raster as IGeoDataset;
        IGeoDataset outputDataset=reclassOp.ReclassByRemap(rasDataset,pRe-
map,true);
        ShowRasterResult(outputDataset,layerName+"Reclass");

        //结果保存到指定位置
        string outputPath=System.IO.Path.GetDirectoryName(outputFileName);
```

```
        string fileName=System.IO.Path.GetFileName(outputFileName);
        IWorkspace ws=OpenRasterWorkspace(outputPath);
        ISaveAs pSaveAs=outputDataset as ISaveAs;
        pSaveAs.SaveAs(fileName,ws,"IMAGINE Image");
    }
```

5. 分类方式变化响应函数 cbxTypeCls_SelectedIndexChanged( )

本函数控制“间隔输入”文本框是否有效，还激活 cbxInLayer_SelectedIndexChanged( )：

```
private void cbxTypeCls_SelectedIndexChanged(object sender,EventArgs e)
{
    string TypeCls = cbxTypeCls.Items[cbxTypeCls.SelectedIndex].ToString();
    if(TypeCls=="自定义")
    {
        this.lblInterval.Visible=true;
        this.txtInterval.Visible=true;
    }
    else
    {
        this.lblInterval.Visible=false;
        this.txtInterval.Visible=false;
    }

    cbxInLayer_SelectedIndexChanged(sender,e);
}
```

13.3.3　核心函数

1. RemapRefresh( )

RemapRefresh( )首先获取分级数组填充 DataTable。然后更新 DataGridView 数据源，并刷新显示。

```
private void RemapRefresh(rstPixelType pixelType)
{
    //获取分级数组
    int breakcount=Convert.ToInt16(cbxCount.Text); //分级数量:
    double[] clsValue=CreateBreakClass(m_iArttributeTable,breakcount);

    //填充新旧值对照表
    if (clsValue!=null)
```

```
        {
            //填充 DataTable
            m_pDataTable.Rows.Clear();
            DataRow FilterRow;
            for (int i=0; i < clsValue.Length - 1; i++)
            {
                FilterRow=m_pDataTable.NewRow();
                if (pixelType!=rstPixelType.PT_DOUBLE && pixelType!=rst-
PixelType.PT_FLOAT)
                {
                    FilterRow[0]=((int)clsValue[i]).ToString()+" - " +
                                    ((int)clsValue[i+1]).ToString();
                }
                else
                {
                    FilterRow[0]=clsValue[i].ToString("F6")+" - " +
                                    clsValue[i+1].ToString("F6");
                }

                FilterRow[1]=i+1;
                m_pDataTable.Rows.Add(FilterRow);
            }

            //刷新 DataGridView
            dataGridTable.DataSource=m_pDataTable;
            dataGridTable.Refresh();
        }
    }
```

2. CreateBreakClass()函数

本函数创建分级数组。方法是从波段数据属性表[Value]字段和[Count]字段获取需要的[值]及其[频数]，以及相应统计信息，然后根据分类方式创建分级器，执行分级计算。

代码如下：

```
    private double[] CreateBreakClass(ITable pTable, int numDesired-
Classes)
    {
        // Specified a field and get the field index for the specified
field to be rendered.
        string valueFieldName="Value"; //Value is the default field,you
```

```
can specify other field here.
        string countFieldName="Count";
        int valueIndex=System.Convert.ToInt32(pTable.FindField(value-
FieldName));
        int countIndex=System.Convert.ToInt32(pTable.FindField(count-
FieldName));
        int NumOfValues=System.Convert.ToInt32(pTable.RowCount(null));

        //从 Table 获取[值]及其[频数]
        int[] dataFrequency=new int[NumOfValues];
        double[] dataValues=new double[NumOfValues];
        IRow pRow=null;
        int i=0;
        ICursor pCursor=pTable.Search(null,false);
        while ( ( pRow=pCursor.NextRow())!=null )
        {
            dataValues[i]=Convert.ToDouble(pRow.get_Value(valueIndex));
            dataFrequency[i]=(countIndex < 0)? 1:Convert.ToInt32(pRow.get_
Value(countIndex));
            i++;
        }

        //创建分级器
        IClassifyGEN pClassify=CreateClassifyGEN( );

        //分级计算
        //pClassify.SetHistogramData(dataValues,dataFrequency)
        pClassify.Classify(dataValues,dataFrequency,ref numDesiredClasses);

        //返回
        return pClassify.ClassBreaks as double[];
    }
```

3. CreateClassifyGEN( )

本函数创建分级器。代码如下：

```
    private IClassifyGEN CreateClassifyGEN( )
    {
        string TypeCls = cbxTypeCls.Items[cbxTypeCls.SelectedIndex].ToS-
tring();
```

```
        double dblCurrentIntervalRange=double.Parse(txtInterval.Text);

        IClassifyGEN pClassify=null;
        switch (TypeCls)
        {
            case "等间距":
                {
                    pClassify=new EqualInterval();
                    break;
                }
            case "分位数":
                {
                    pClassify=new Quantile();
                    break;
                }
            case "自然分割":
                {
                    pClassify=new NaturalBreaks();
                    break;
                }
            case "几何间隔":
                {
                    pClassify=new GeometricalIntervalClass();
                    break;
                }
            case "标准差":
                {
                    pClassify=new StandardDeviation();
                    IDeviationInterval pStdDev=(IDeviationInterval)pClas-
sify;
                    pStdDev.Mean=m_rstStatistic.Mean;
                    pStdDev.StandardDev=m_rstStatistic.StandardDeviation;
                    pStdDev.DeviationInterval=1;
                    break;
                }
            case "自定义":
                {
                    pClassify=new DefinedInterval();
```

```
                    IIntervalRange pIntervalRange = (IIntervalRange)pClas-
sify;
                    pIntervalRange.IntervalRange=dblCurrentIntervalRange;
                    break;
                }
            default:
                {
                    break;
                }
        }
        return pClassify;
    }
```

4. getAttrTableByErgodic( )

对于离散栅格数据直接返回其属性表即可。对于浮点类栅格数据，其属性表不存在。采用如下思路构造属性表：

①则按小数点后 6 位的准确度离散化；

②Dictionary 结构统计频数；

③Dictionary 转换为 ITable。

代码如下：

```
    private ITable getAttrTableByErgodic(IRaster pRaster)
    {
        IRasterProps rasterProps=(IRasterProps)pRaster;
        IRaster2 pRaster2=pRaster as IRaster2;
        int nodata=0;
        //遍历像元阵列,统计准确到 0.000001 的像元频数
        Dictionary<double,long> ValueFrequence=new Dictionary<double,
long>();
        for (int y=0; y<rasterProps.Height; y++)
        {
            for (int x=0; x < rasterProps.Width; x++)
            {
                //object obj=pSafeArray.GetValue(x,y);
                object obj=pRaster2.GetPixelValue(0,x,y);
                if (obj==null)
                {
                    nodata++;
                    continue;
                }
```

```
                double value=Convert.ToDouble(obj);
                value=Math.Round(value*1000000.0)/ 1000000.0;
                if (!ValueFrequence.ContainsKey(value))
                {
                    ValueFrequence.Add(value,1);
                }
                else if (ValueFrequence.ContainsKey(value))
                {
                    ValueFrequence[value]=ValueFrequence[value]+1;
                }
            }
        }

        //Dictionary 转 ITable
        ITable pTable=InitializeITable();
        IOrderedEnumerable<KeyValuePair<double,long>> sort=
                                    ValueFrequence.OrderBy(x=> x.Key);
        foreach (KeyValuePair<double,long> kv in sort)
        {
            IRow row=pTable.CreateRow();
            row.Value[0]=kv.Key;
            row.Value[1]=kv.Value;
            row.Store();
        }
        return pTable;
    }
```

5. ConstructRemap( )

ConstructRemap( ) 根据 DataGridView 新旧对照数据构造一个映射表，代码如下：

```
private IRemap ConstructRemap()
{
    INumberRemap pSRemap=new NumberRemapClass();
    for (int i=0; i < dataGridTable.RowCount-1; i++)
    {
        String str=dataGridTable[0,i].Value.ToString();

        float fvalue,tvalue;
        int p=str.LastIndexOf(" - ");
```

```
            fvalue=Convert.ToSingle(str.Substring(0,p));
            tvalue=Convert.ToSingle(str.Substring(p+3,str.Length - (p+3)));
            pSRemap.MapRange(fvalue,tvalue,i+1);
        }

        return (IRemap)pSRemap;
    }
```

13.3.4 辅助函数

代码如下：

```
    //内存中初始化一个内存 ITable
    private ITable InitializeITable()
    {
        IWorkspaceFactory workspaceFactory = new InMemoryWorkspaceFac-
toryClass();
        IWorkspaceName workspaceName = workspaceFactory.Create("","My-
Workspace",null,0);
        ESRI.ArcGIS.esriSystem.IName name=(IName)workspaceName;
        ESRI.ArcGIS.Geodatabase.IWorkspace inmemWor=(IWorkspace)name.Open();

        IFields fields=new FieldsClass();
        IFieldsEdit fieldsEdit=(IFieldsEdit)fields;
        fieldsEdit.FieldCount_2=2;

        IFieldEdit col=new FieldClass();
        col.Name_2="Value";
        col.Type_2=esriFieldType.esriFieldTypeDouble;
        fieldsEdit.set_Field(0,col);

        col=new FieldClass();
        col.Name_2="Count";
        col.Type_2=esriFieldType.esriFieldTypeInteger;
        fieldsEdit.set_Field(1,col);

        ESRI.ArcGIS.Geodatabase.ITable pTable=null;
        pTable=(inmemWor as IFeatureWorkspace).CreateTable("重分类",
fields,null,null,"");
        return pTable;
```

```
    }

    //指定路径的工作空间
    private IWorkspace OpenRasterWorkspace(string outputPath)
    {
        IWorkspaceFactory wsf=new RasterWorkspaceFactoryClass();
        IWorkspace ws=wsf.OpenFromFile(outputPath,0);
        //rasAnaEnv.OutWorkspace=ws;
        return ws;
    }

    //获取当前视图中栅格图层集合
    private IEnumLayer GetRasterLayers()
    {
        UID uid=new UIDClass();
        //uid.Value="{40A9E885-5533-11d0-98BE-00805F7CED21}";//Featu-
reLayer
        uid.Value="{D02371C7-35F7-11D2-B1F2-00C04F8EDEFF}"; //Raster-
Layer
        IEnumLayer layers=m_mapControl.Map.get_Layers(uid,true);

        return layers;
    }

    //通过图层名得到栅格图层
    private IRasterLayer GetRasterLayer(string layerName)
    {
        //get the layers from the maps
        IEnumLayer layers=GetRasterLayers();
        layers.Reset();

        ILayer layer=null;
        while ((layer=layers.Next())!=null)
        {
            if (layer.Name==layerName)
                return layer as IRasterLayer;
        }
        return null;
```

```
    }

    //显示栅格结果
    private void ShowRasterResult(IGeoDataset geoDataset,string inter-
Type)
    {
        IRasterLayer rasterLayer=new RasterLayerClass();
        IRaster raster=new Raster();
        raster=(IRaster)geoDataset;
        rasterLayer.CreateFromRaster(raster);
        rasterLayer.Name=interType;

        m_mapControl.AddLayer((ILayer)rasterLayer,0);
        m_mapControl.ActiveView.Refresh();
    }
```

13.4 功能调用

在【Spatial Analysis】页上添加【Reclassify】按钮。建立 Click 响应函数;

```
private void btnReclassify_Click(object sender,EventArgs e)
{
ReclassifyFrm frm=new Reclassify (_mapControl);
     frm.Show();
}
```

13.5 编译测试

按下 F5 键，编译运行程序，点击菜单“Reclassify”，弹出分析窗口，添加分析图层，并设置输出文件路径和文件名，确认映射表是否符合要求(必要时做适当编辑)，按【分析】按钮即生成重新分类后的数据。

第 14 章　成本路径分析

14.1　知识要点

距离分析是指根据每一栅格相距其最邻近“源”的距离计算，得到每一栅格与其邻近目标“源”的相互关系。进行距离分析时，如果考虑通过每一个栅格的通行成本(时间、金钱等)，即是所谓成本距离，否则就是“欧氏距离”分析。成本数据通常是一个单独的成本栅格数据(一般基于重分类来完成)。

在 ArcGIS Engine 中，RasterDistanceOpClass 类实现了距离分析。该类实现了两个主要的接口，分别是 IDistanceOp 和 IRasterAnalysisEnvironment 接口。IDistanceOp 接口包含了距离分析的所有方法，主要有：

①EucDistance：欧氏距离；

②EucDirection：欧氏方向；

③EucAllocation：欧氏分配；

④CostDistance：成本距离；

⑤CostBackLink：成本回溯链接；

⑥CostAllocation：成本分配；

⑦CostPath：成本路径；

⑧Corridor：廊道分析。

14.2　功能描述

单击【Saptial Analysis】页上【Cost Distance】按钮，弹出成本距离分析对话框，如图 14-1 所示，即可根据输入成本图层、源图层、目标图层生成成本路径栅格数据集。

图 14-1　成本距离分析对话框

14.3　功能实现

14.3.1　新建功能窗体

1. 界面设计

项目中添加一个新的窗体，名称为“CostDistanceFrm”，Name 属性设置为“成本距离”，添加 3 个 ComboBox、1 个 TextBox、2 个 Button 控件。

控件属性设置见表 14-1。

表 14-1　**控件及其属性、说明**

控件类型	Name 属性	控件说明	备　注
ComBox	cbxCostLayer	成本图层:	
ComBox	cbxSourceLayer	源图层:	
ComBox	cbxTargetLayer	目标图层:	
TextBox	txtOutLayer	输出结果文件名	
Button	btnAnalyst	分析按钮	
Button	btnCancel	取消按钮	

2. 类结构设计

添加如下引用代码，修改类定义代码：

```
public partial class CostDistanceFrm:Form
```

```
{
    private IMapControl3 m_mapControl;
    public CostDistanceFrm(IMapControl3mapControl)
    {
        InitializeComponent();
        m_mapControl =mapControl;
    }

    //窗体加载时触发事件,执行函数
    private void CostDistanceFrm_Load(object sender,EventArgs e)
    //点击输出路径按钮时,执行函数
    private void btnOutLayer_Click(object sender,EventArgs e)
    //点击响应函数,执行操作
    private void btnAnalyst_Click(object sender,EventArgs e)
    //取消按钮响应函数:
    private void BtnCancel_Click(object sender,EventArgs e)

    //若干功能函数:
    public void CalCostDistance(IRasterLayer costRasterLayer,IFeatureClass sourceFclass,IFeatureClass targetFclass,string outputFileName)
    private void SetAnalysisEnvironment(IRasterAnalysisEnvironment rasAnaEnv,IRaster costRaster)
    ……
}
```

14.3.2 消息响应函数

1. 载入响应函数 CostDistanceFrm_Load()

①用 m_ mapControl 栅格图层名填充 cbxCostLayers；矢量图层名填充 cbxSourceFLayer、cbxTargetFLayer；

②设置输出文件的默认输出路径，这里我们将默认输出路径设置为系统临时目录。

代码如下：

```
private void CostDistanceFrm_Load(object sender,EventArgs e)
{
    //填充 cbxCostLayer
    for (int i=0; i < m_mapControl.LayerCount; i++)
    {
        ILayer pLayer;
        pLayer=m_mapControl.get_Layer(i);
```

```
        if (pLayer is IRasterLayer)
        {
            cbxCostLayer.Items.Add(pLayer.Name);
        }
    }

    //填充 cbxSourceFLayer、cbxTargetFLayer
    for (int i=0; i < m_mapControl.LayerCount; i++)
    {
        ILayer pLayer;
        pLayer=m_mapControl.get_Layer(i);
        if (pLayer is IFeatureLayer)
        {
            cbxSourceFLayer.Items.Add(pLayer.Name);
            cbxTargetFLayer.Items.Add(pLayer.Name);
        }
    }

    this.txtOutLayer.Text=@"C:\Temporary2015\Temp.FIFF";
}
```

2. 输出路径设置响应函数 btnOutLayer_ Click()

输出路径设置由 SaveFileDialog 实现，添加代码如下：

```
private void btnOutLayer_Click(object sender,EventArgs e)
{
    SaveFileDialog saveDlg=new SaveFileDialog();
    saveDlg.CheckPathExists=true;
    saveDlg.Filter="Rasterfile (*.img) |*.img";
    saveDlg.OverwritePrompt=true;
    saveDlg.Title="Output Layer";
    saveDlg.RestoreDirectory=true;
    saveDlg.FileName=(string)cbxInLayer.SelectedItem+"_Reclass.img";

    DialogResult dr=saveDlg.ShowDialog();
    if (dr==DialogResult.OK)
        txtOutLayer.Text=saveDlg.FileName;
}
```

3. 分析响应函数 btnAnalyst_Click()

①准备参数：源、目标、成本；

②调用 CalCostDistance( )；

代码如下：

```
private void btnAnalyst_Click(object sender,EventArgs e)
{
    string outputFileName=this.txtOutLayer.Text;
    string layerName=this.cbxCostLayer.SelectedItem.ToString();
    IRasterLayer costRasterLayer=getLayerFromName(layerName) as IRas-
terLayer;

    layerName=this.cbxSourceFLayer.SelectedItem.ToString();
    IFeatureLayer sourceLayer=getLayerFromName(layerName) as IFea-
tureLayer;

    layerName=this.cbxTargetFLayer.SelectedItem.ToString();
    IFeatureLayer targetLayer=getLayerFromName(layerName) as IFea-
tureLayer;

    CalCostDistance(costRasterLayer,sourceLayer.FeatureClass,tar-
getLayer.FeatureClass,outputFileName);
}
```

14.3.3 核心函数

1. CalCostDistance( )

CalCostDistance( ) 完成成本路径分析全部工作，步骤如下：其中步骤③至④是为⑤准备数据。

①创建距离分析接口对象；

②设置分析环境；使用 SetAnalysisEnvironment( ) 函数；

③成本距离计算；调用 IDistanceOp2 接口的 CostDistance 方法，输入源数据集(可以是点类型，也可是线类型)、成本数据集，生成由源到达分析区域内任意一点的成本距离。

④回溯链接方向计算；调用 IDistanceOp2 接口的 CostBackLink 方法，输入源数据集、成本数据集，生成由源到达分析区域内任意一点的回溯链接方向。

⑤成本路径计算；调用 IDistanceOp2 接口的 CostPath 方法，输入成本数据集、回溯链接方向数据集、目标数据集(只能是点类型，否则结果无意义)，生成由源到目标的最短成本路径。

⑥结果保存到指定位置；

```
public void CalCostDistance(IRasterLayer costRasterLayer,IFeature-
Class sourceFclass,
    IFeatureClass targetFclass,string outputFileName)
```

```
    {
        IGeoDataset sourceDs=sourceFclass as IGeoDataset;
        IGeoDataset targerDs=targetFclass as IGeoDataset;
        IGeoDataset costDs=costRasterLayer.Raster as IGeoDataset;

        //创建距离分析接口对象
        IDistanceOp2 distanceOp=new RasterDistanceOpClass();
        //设置分析环境
        IRasterAnalysisEnvironment rasAnaEnv=distanceOp as IRasterAnalys-
isEnvironment;
        SetAnalysisEnvironment(rasAnaEnv,costRasterLayer.Raster);

        //成本距离计算
        object maxDistance=System.Reflection.Missing.Value;
        object valueRaster=System.Reflection.Missing.Value;
        IGeoDataset distanceDs=distanceOp.CostDistance(sourceDs,costDs,
ref maxDistance,ref valueRaster);
        ShowRasterResult(distanceDs,"CostDistance");

        //回溯链接方向计算
        IGeoDataset backLink=distanceOp.CostBackLink(sourceDs,costDs,
ref maxDistance,ref valueRaster);
        ShowRasterResult(backLink,"BackLink");

        //成本路径计算
        IGeoDataset outputDataset=distanceOp.CostPath(targerDs,distanceDs,
backLink,esriGeoAnalysisPathEnum.esriGeoAnalysisPathBestSingle);
        ShowRasterResult(outputDataset,"CostPath");

        //结果保存到指定位置
        IWorkspace ws=OpenRasterWorkspace(outputFileName);
        string fileName=System.IO.Path.GetFileName(outputFileName);
        ISaveAs pSaveAs=outputDataset as ISaveAs;
        pSaveAs.SaveAs(fileName,ws,"IMAGINE Image");
    }
```

2. SetAnalysisEnvironment()函数

本函数负责设置栅格分析环境，这是十分重要的一步，所有基于栅格分析的分析方法，都必须设置好分析环境，它是通过 IRasterAnalysisEnvironment 接口来实现的，所有进行栅格分析的组件都实现了 IRasterAnalysisEnvironment 接口。

代码如下:

```
//分析环境设置
private void SetAnalysisEnvironment ( IRasterAnalysisEnvironment rasAnaEnv,
                                       IRaster pRaster)
{
    //设置生成图层的范围
    IGeoDataset rGeoDataset=pRaster as IGeoDataset;
    object extent=rGeoDataset.Extent;
    object missing=System.Reflection.Missing.Value;
    rasAnaEnv.SetExtent(esriRasterEnvSettingEnum.esriRasterEnvValue, ref extent,ref missing);

    //设置生成图层的栅格大小
    IRasterProps rProps=pRaster as IRasterProps;
    IPnt p=rProps.MeanCellSize();
    object cellsize=(p.X+p.Y)/2;
    rasAnaEnv.SetCellSize(esriRasterEnvSettingEnum.esriRasterEnvMinOf, ref cellsize);
}
```

14.3.4 辅助函数

```
//生成图层的工作空间
private IWorkspace OpenRasterWorkspace(string outputFileName)
{
    IWorkspaceFactory wsf=new RasterWorkspaceFactoryClass();
    string outputPath=System.IO.Path.GetDirectoryName(outputFileName);

    IWorkspace ws=wsf.OpenFromFile(outputPath,0);
    return ws;
}

//显示栅格结果
private void ShowRasterResult(IGeoDataset geoDataset,string interType)
{
    IRasterLayer rasterLayer=new RasterLayerClass();
```

```
    IRaster raster=new Raster();
    raster=(IRaster)geoDataset;
    rasterLayer.CreateFromRaster(raster);
    rasterLayer.Name=interType;

    m_mapControl.AddLayer((ILayer)rasterLayer,0);
    m_mapControl.ActiveView.Refresh();
}
```

14.4 功能调用

在【Saptial Analysis】Tab 页，添加【Cost Distance】按钮。建立 Click 响应函数；

```
private void btnCostDistance_Click(object sender,EventArgs e)
{
    CostDistabceFrm frm=new CostDistabceFrm (_mapControl);
     frm.Show();
}
```

14.5 编译测试

按下 F5 键，编译运行程序，点击【Cost Distance】按钮，弹出分析窗口，添加分析图层，并设置输出文件路径和文件名。

第 15 章　运输网络分析

15.1　知识要点

GIS 网络是由一系列相互连通的点和线组成，用来描述地理要素(资源)的流动情况的特殊数据结构。其中网络边是具有一定长度和物流的网络要素，节点是两条或两条以上边的交会处，是两条边之间进行物流转换的网络要素。

GIS 的网络分析是依据网络的拓扑关系，通过考察网络元素的空间及属性数据，以图论和运筹学等数学理论模型为基础，对网络的性能特征进行多方面评价的一种分析计算。

ArcGIS Engine 网络分析有两种方式，一是网络数据集(网络数据结构为 NetworkDataset)，网络中流动的资源自身可以决定流向(一些特别限制除外)；二是几何网络(网络数据结构为 Geometric Network)，网络中流动的资源自身不能决定流向（如水流、电流)。

本章介绍基于 NetworkDataset 的网络分析方法(为简化学习曲线，网络结构数据由 ArcCatalog 创建)，主要接口为 INASolver、INAContext。

网路分析的思路分为以下几步：

①打开网络数据集；

②创建网络分析求解器 INASolver 接口对象，实现该接口的类分别是：

◆ NARouteSolver(最短路径分析)；

◆ NAServiceAreaSolver(服务范围分析)；

◆ NAClosestFacilitySolver(最近服务设施分析)；

◆ NAODCostMatrixSolverClass(成本矩阵分析)；

◆ NALocationAllocationSolverClass(位置指派分析)。

③创建网络分析上下文对象 INAContext，该接口主要作用是将 INASolver、INetworkDataset、INAClasses(位置数据类)等对象集成在一起，构成网络分析环境。

该接口用 INASolver 接口 CreateContext()方法创建，然后转换为 INAContextEdit 接口捆绑网络数据集。

④为 INAContext 加载位置点信息；用到 INAClassLoader。

⑤设置分析参数；用到 INASolverSettings 接口，INASolver 转换得到该接口。

⑥进行分析；用 INASolver 的 Solve()函数。

⑦显示结果信息。

15.2　功能描述

点击【Network】Tab 页，如果网络数据集已经加载到视图(可用 Add Data 工具)，则 RibbonBar：【Network Dataset Analysis】所有按钮有效，如图 15-1 所示。

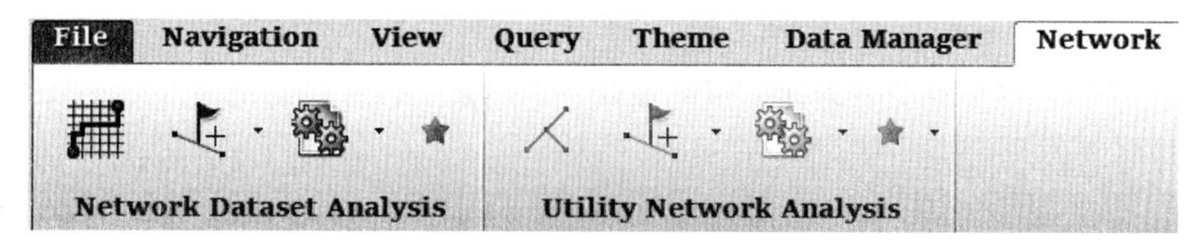

图 15-1　网络数据集分析按钮

包括 4 个按钮(从左到右顺序)，作用如下：

①求解(btnNetworkDatasetSolve)按钮：分析求解。

②位置添加(btnNetworkAddPosition)下拉菜单(ButtonItem 类型)，如图 15-2(a)：添加站点(Add Stops)、添加障碍(Add Barriers)、添加设施(Add Facility)、添加事故(Add Incidents)。

③方法选择下拉菜单(CheckBoxItem 类型)，如图 15-2(b)所示：Route，Service Area，Closest Facility。

④清除按钮(Clear)，清除所有的位置点。

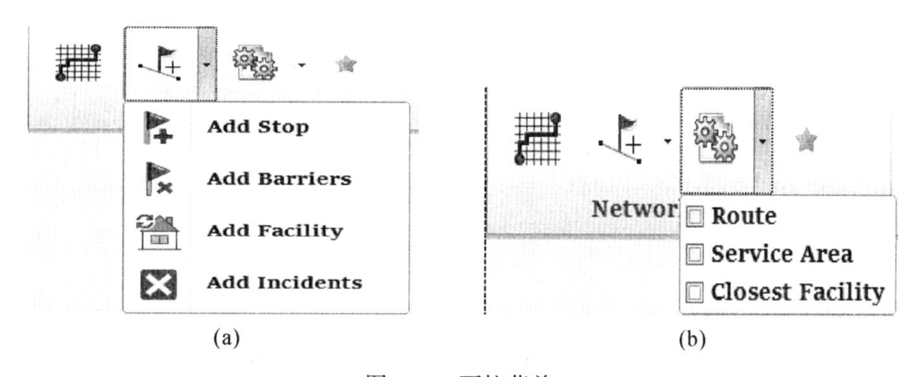

(a)　　(b)

图 15-2　下拉菜单

15.3　功能实现

15.3.1　工具条功能实现

1. 站点/障碍点添加功能

站点/障碍点添加功能激活后，光标在屏幕上拾取站点/障碍点的位置，将其转化为

Mark 类型的图形元素，显示站点/障碍点符号标志；同时将其转化为 Point 要素对象分别存入临时数据库(TemporaryGeodatabase.gdb)的 Stops，Barriers 要素类中，供之后分析用。

功能类继承 BaseTool 实现，在新建项的模板浏览窗口，选择 ArcEngine 的 BaseTool 模板，功能类取名为 AddNetStopsTool/AddNetBarriesTool。新工具框架生成后，重载鼠标响应函数：OnClick()、OnMouseDown()。

AddNetStopsTool 关键源代码如下(AddNetBarriesTool 类似)：

```
//定义私有成员
private IHookHelper m_hookHelper=null;
private IFeatureClass inputFClass;
string path = System.AppDomain.CurrentDomain.SetupInformation.ApplicationBase;
//重载点击按钮事件函数:打开 TemporaryGeodatabase.gdb 中 Stops 要素类,并清空
public override void OnClick()
{
    // TODO:Add AddNetStopsTool.OnClick implementation
    string strName=NetWorkAnalysClass.getTemporaryPath(path)+
                                  "\\data\\TemporaryGeodatabase.gdb";
    IFeatureWorkspace pFWorkspace=OpenWorkspace(strName) as IFeatureWorkspace;
    inputFClass=pFWorkspace.OpenFeatureClass(LocationTypeEnum.Stops.ToString());
    if (inputFClass.FeatureCount(null) >0 )
    {
        ITable pTable=inputFClass as ITable;
        pTable.DeleteSearchedRows(null);
    }
}
//重载鼠标落下函数
public override void OnMouseDown(int Button,int Shift,int X,int Y)
{
    try
    {
        //将屏幕点转换为 PointClass
        IPoint pStopsPoint=new PointClass();
        pStopsPoint=
                m_hookHelper.ActiveView.ScreenDisplay.DisplayTransformation.ToMapPoint(X,Y);
```

```
            //创建一个点要素,将来分析之用
            IFeature newPointFeature=inputFClass.CreateFeature();
            newPointFeature.Shape=pStopsPoint;
            newPointFeature.Store();

            //创建一个图形元素,显示停靠点效果
            IMarkerElement pMarkerEle=new MarkerElementClass();
            string picturePath = NetWorkAnalysClass.getPath(path) +" \\
data \\Img \\stops.bmp";
            pMarkerEle.Symbol =CreatePictureMarkerSymbol( picturePath );
            IElement pEle=pMarkerEle as IElement;
            pEle.Geometry=pStopsPoint;

            //图形元素添加到 IGraphicsContainer
            IGraphicsContainer pGrap=m_hookHelper.ActiveView as IGraphic-
sContainer;
            pGrap.AddElement(pEle,1);

            //刷新图形元素
            m_hookHelper.ActiveView.PartialRefresh(esriViewDrawPhase.
esriViewGraphics,null,null);
        }
        catch
        {
            MessageBox.Show("添加站点失败!","提示",MessageBoxButtons.OK,
                          MessageBoxIcon.Warning);
            return;
        }
    }
```

设施点/事故点的添加按钮功能，完全类似于站点/障碍点添加功能，不同的是存入到临时数据库的不同的数据集 Facilities、Incidents 中。

2. 分析(Solve)功能

Solve 功能类继承 BaseCommand 实现，在新建项的模板浏览窗口，选择 ArcEngine 的 BaseCommand 模板，功能类取名为 SolveTool。新工具框架生成后，重载鼠标响应函数 OnClick()；本函数根据 m_ NASelectedString 变量值分别调用不同的分析函数。

OnClick() 源代码如下：

```
//添加类成员变量：
private IHookHelper m_hookHelper=null;
```

```
    private INetworkDataset m_NetworkDataset = null;
    private string m_NASelectedString = "";
    string path = System.AppDomain.CurrentDomain.SetupInformation.Ap-
plicationBase;

    //修改构造函数
    public SolveTool (INetworkDataset pNetworkDataset, string strNASe-
lected)
    {
        m_NetworkDataset = pNetworkDataset;
        m_NASelectedString = strNASelected;

        base.m_category = "NetWorkAnalyst"; //localizable text
        base.m_caption = "Solve";  //localizable text
        base.m_message = "Solve";  //localizable text
        base.m_toolTip = "Solve";
        base.m_name = "SolveTool";
        try
        {
            string bitmapResourceName = GetType().Name+".bmp";
            base.m_bitmap = new Bitmap(GetType(),bitmapResourceName);
        }
        catch (Exception ex)
        {
             System.Diagnostics.Trace.WriteLine(ex.Message," Invalid
Bitmap");
        }
    }

    public override void OnClick()
    {
        string strName = NetworkAnalyst.getTemporaryPath(path) +"\\data
\\HuanbaoGeodatabase.gdb";
        IFeatureWorkspace pFWorkspace = NetworkAnalyst.OpenWorkspace(str-
Name) as IFeatureWorkspace;
        switch (m_NASelectedString)
        {
            case "Route":
```

```
                shortestRouteAnalyst(pFWorkspace);
                break;
            case "ServiceArea":
                ServiceAreaAnalyst(pFWorkspace);
                break;
            case "ClosestFacility":
                ClosestFacilityAnalyst(pFWorkspace);
                break;
            default:
                break;
        }
    }
```

分析函数步骤如下：

①创建网络分析上下文对象；

②加载站点/障碍点要素，并设置捕捉容差；

③设置分析参数；

④执行分析操作；

⑤创建 INALayer 层，显示分析结果。

这里用到 NetworkAnalyst 核心功能类，后文将介绍。

```
//最近设施分析函数
private void ClosestFacilityAnalyst(IFeatureWorkspace pFWorkspace)
{
    //创建网络分析上下文
    INAContext pNAContext = NetworkAnalyst.CreateSolverContext(m_
NetworkDataset,"ClosestFacility");

    //加载设施点并设置容差
    IFeatureClass facilitiesFClass =
                pFWorkspace.OpenFeatureClass(LocationTypeEnum.
Facilities.ToString());
    IFeatureClass incidentsFClass =
                pFWorkspace.OpenFeatureClass(LocationTypeEnum.
Incidents.ToString());
    NetworkAnalyst.LoadNANetworkLocations(pNAContext, Location-
TypeEnum.Facilities.ToString(),facilitiesFClass as ITable,80);
    NetworkAnalyst.LoadNANetworkLocations(pNAContext, Location-
TypeEnum.Incidents.ToString(),incidentsFClass as ITable,80);
```

```
        //分析参数设置
        string ImpedanceName = getImpedanceAttributeName(m_NetworkData-
set);
        NetworkAnalyst. SetClosestFacilitySolverSettings ( pNAContext,
ImpedanceName,2);

        //执行分析操作
        NetworkAnalyst.Solve(pNAContext);

        //创建 INALayer 层
        INALayer naLayer = pNAContext.Solver.CreateLayer(pNAContext);
        ILayer layer1 = naLayer.get_LayerByNAClassName("CFRoutes");
        layer1.Name = "最近设施";

        m_hookHelper.FocusMap.AddLayer(layer1);
        m_hookHelper.ActiveView.Refresh();
    }

    //最近设施分析函数
    private void ServiceAreaAnalyst(IFeatureWorkspace pFWorkspace)
    {
        //创建网络分析上下文
        INAContext pNAContext = NetworkAnalyst.CreateSolverContext ( m_
NetworkDataset,"ServiceArea");

        //加载设施点并设置容差
        IFeatureClass facilitiesFClass =
                        pFWorkspace. OpenFeatureClass ( LocationTypeEnum.
Facilities.ToString());
        NetworkAnalyst. LoadNANetworkLocations ( pNAContext, Location-
TypeEnum.Facilities.ToString(),facilitiesFClass as ITable,80);

        //分析参数设置
        IDoubleArray pBrks = new DoubleArrayClass();
        {
            pBrks.Add(1000);
            pBrks.Add(2000);
            pBrks.Add(3000);
```

```
        }
        string ImpedanceName=getImpedanceAttributeName(m_NetworkDataset);
        NetworkAnalyst.SetServiceAreaSolverSettings(pNAContext,Imped-
anceName,pBrks);

        //执行分析操作
        NetworkAnalyst.Solve(pNAContext);

        //创建 INALayer 层
        INALayer naLayer=pNAContext.Solver.CreateLayer(pNAContext);
        ILayer layer1=naLayer.get_LayerByNAClassName("SAPolygons");
        layer1.Name="服务范围";

        m_hookHelper.FocusMap.AddLayer(layer1);
        m_hookHelper.ActiveView.Refresh();
    }

    //最短路径分析函数
    private void shortestRouteAnalyst(IFeatureWorkspace pFWorkspace)
    {
        //创建网络分析上下文
        INAContext pNAContext = NetworkAnalyst.CreateSolverContext(m_
NetworkDataset,"Route");

        //加载站点/障碍点要素,并设置容差
        IFeatureClass stopsFClass=
                        pFWorkspace. OpenFeatureClass(LocationTypeEnum.
Stops.ToString());
        IFeatureClass barriesFClass=
                        pFWorkspace. OpenFeatureClass(LocationTypeEnum.
Barriers.ToString());
        NetworkAnalyst.LoadNANetworkLocations(pNAContext,
                        LocationTypeEnum. Stops. ToString(), stopsFClass
as ITable,80);
        NetworkAnalyst.LoadNANetworkLocations(pNAContext,
                        LocationTypeEnum.Barriers. ToString(),barriesF-
Class as ITable,5);
```

```
        //分析参数设置
        string ImpedanceName = getImpedanceAttributeName(m_NetworkData-
set);
        NetworkAnalyst. SetRouteSolverSettings ( pNAContext, Impedance-
Name,true);

        //执行分析操作
        NetworkAnalyst.Solve(pNAContext);

        //创建 INALayer 层
        INALayer naLayer = pNAContext.Solver.CreateLayer(pNAContext);
        ILayer layer1 = naLayer.get_LayerByNAClassName("Routes");
        layer1.Name = "最短路径";

        m_hookHelper.FocusMap.AddLayer(layer1);
        m_hookHelper.ActiveView.Refresh();
    }

    //获取代价属性名(这里网络定义中的第一个)
    private string getImpedanceAttributeName(INetworkDataset pNetwork-
Dataset)
    {
        //设置路径分析阻抗属性(Set the impedance attribute)
        for (int i = 0; i < pNetworkDataset.AttributeCount; i++)
        {
             INetworkAttribute networkAttribute = pNetworkDataset.get_
Attribute(i);
            if (networkAttribute.UsageType == esriNetworkAttributeUsageType.
esriNAUTCost)
            {
                return networkAttribute.Name;
            }
        }

        return "";
    }
```

15.3.2 核心功能类的实现

为方便操作，将网络分析的核心功能封装为 NetWorkAnalyst 类，共有成员为静态成员，方便直接调用。

1. NetWorkAnalyst 设计

代码如下：

```
class NetWorkAnalysClass
{
    //载入位置数据,Stops 或 Barries
     public static void LoadNANetworkLocations ( LocationTypeEnum
enumLocation, ITableinputFC, INAContext pNAContext, double dSnapToler-
ance)

    //创建路径分析上下文 INAContext
     public static INAContext CreateSolverContext ( INetworkDataset
networkDataset,string type)

    // Solve the problem 进行最短路径分析
    public static string Solve(INAContext pNAContext)

    // Set Solver Settings 设置分析参数(阻抗属性等)
    public static void SetRouteSolverSettings ( INAContext NAContext,
string ImpedanceName,bool is-RouteOptimal)
    public static void SetClosestFacilitySolverSettings(INAContext
NAContext,string ImpedanceName,int targetCount)
    public static void SetServiceAreaSolverSettings(INAContext NACon-
text,string ImpedanceName,IDoubleArray BreaksArr)

    //获取临时数据库存放路径
    public static string getTemporaryPath(string path)

    //若干辅助函数
    ……
}
```

这里定义枚举类型 LocationTypeEnum，且要求临时数据库中相应的要素类名称与枚举名称要相同。

```
public enum LocationTypeEnum
```

```
{
    Stops=0,
    Barriers=1,
    Facilities=2,
    Incidents=3,
    Origins=4,
    Destinations=5,
}
```

2. NetWorkAnalyst 实现

(1)CreateSolverContext()

本函数根据已经打开的网络数据集，创建路径分析上下文 INAContext，源代码如下：

```
public static INAContext CreateSolverContext(INetworkDataset net-
workDataset,string type)
{
    INASolver naSolver;
    switch(type)
    {
        case "Route":
            naSolver=new NARouteSolver();
            break;
        case "ServiceArea":
            naSolver=new NAServiceAreaSolverClass();
            break;
        case "ClosestFacility":
            naSolver=new NAClosestFacilitySolverClass();
            break;
        case "ODCostMatrixSolver":
            naSolver=new NAODCostMatrixSolverClass();
            break;
        case "VRPSolver":
            naSolver=new NAVRPSolverClass();
            break;
        case "LocationAllocationSolver":
            naSolver=new NALocationAllocationSolverClass();
            break;
        default:
            naSolver=new NAClosestFacilitySolver();
            break;
```

```
        }

        IDatasetComponent dsComponent =networkDataset as IDatasetCompo-
nent;
        IDENetworkDataset deNDS =dsComponent.DataElement as IDENetwork-
Dataset;

        INAContextEdit contextEdit =null;
        contextEdit =naSolver.CreateContext (deNDS, naSolver.Name) as INA-
ContextEdit;
        contextEdit.Bind(networkDataset,new GPMessagesClass());

        return (contextEdit as INAContext);
    }
```

(2) LoadNANetworkLocations

本函数将记录在临时数据库的位置数据(Stops 或 Barriers)载入 INAContext，步骤如下：

①清空分析上下文中已存在的位置点；

②创建位置加载器 NAClassLoader；

③设置捕捉容限值；

④设置字段匹配；

⑤设置排除网络受限；

⑥加载网络位置点数据；

⑦发送消息。

源代码如下：

```
    public static void LoadNANetworkLocations (INAContext NAContext,
string strNAClassName,ITable inputTable,double dSnapTolerance)
    {
        INamedSet pNamedSet =NAContext.NAClasses;
        INAClass pNAClass =pNamedSet.get_ItemByName(strNAClassName) as
INAClass;

        //删除分析上下文中已存在的位置点
        pNAClass.DeleteAllRows();

        //创建 NAClassLoader
        INAClassLoader loader =new NAClassLoaderClass();
```

```
        {
            loader.Locator=NAContext.Locator;
             loader.Locator.SnapTolerance = dSnapTolerance;//设置捕捉容
限值
            loader.NAClass=pNAClass;

            INAClassFieldMap fieldMap=new NAClassFieldMapClass();
            //fieldMap.set_MappedField("OBJECTID","OBJECTID");
            //fieldMap.set_MappedField("FID","FID");
            fieldMap.CreateMapping(pNAClass.ClassDefinition,inputTable.
Fields);
            loader.FieldMap=fieldMap;
        }

        //排除网络受限
        INALocator3 locator=NAContext.Locator as INALocator3;
        locator.ExcludeRestrictedElements=false;
        locator.CacheRestrictedElements(NAContext);

        //加载网络位置点数据
        int rowsIn=0;
        int rowsLocated=0;
        ICursor pCursor=inputTable.Search(null,true);
        loader.Load(pCursor,null,ref rowsIn,ref rowsLocated);

        //发送消息
        ((INAContextEdit)NAContext).ContextChanged();
    }
```

(3)分析参数设置

代码如下:

```
    //最短路径分析参数设置
    public static void SetRouteSolverSettings(INAContext NAContext,
string ImpedanceName,bool isRouteOptimal)
    {
        INASolver naSolver=NAContext.Solver;

        //Set specific Settings
        INARouteSolver cfSolver=naSolver as INARouteSolver;
```

```
        {
            //设置生成线的类型(基于真实网络的几何形状生成线,并尽可能增加测度)
            cfSolver.OutputLines = esriNAOutputLineType.esriNAOutputLi-
neTrueShapeWithMeasure;
            //创建遍历结果
            cfSolver.CreateTraversalResult = true;
            //不使用时间窗口
            cfSolver.UseTimeWindows = false;
            //线路优化(reorder for optimal route)
            cfSolver.FindBestSequence = isRouteOptimal;
            cfSolver.PreserveFirstStop = false;
            cfSolver.PreserveLastStop = false;
        }

        //Generic Settings
        INASolverSettings naSolverSettings = naSolver as INASolverSet-
tings;
        {
            //设置路径分析阻抗属性(Set the impedance attribute)
            naSolverSettings.ImpedanceAttributeName = ImpedanceName;

            ////设置单行限制(Set the OneWay Restriction if necessary)
            //IStringArray restrictions = naSolverSettings.Restriction-
AttributeNames;
            //restrictions.RemoveAll();
            //restrictions.Add("oneway");
            //naSolverSettings.RestrictionAttributeNames = restrictions;

            //限制 U 形转向限制(Restrict UTurns)
             naSolverSettings.RestrictUTurns = esriNetworkForwardStar-
Backtrack.esriNFSBNoBacktrack;
            //设置忽略无效位置
            naSolverSettings.IgnoreInvalidLocations = true;

            ////设置层次属性(Set the Hierachy attribute)
            //naSolverSettings.UseHierarchy = true;
            //if (naSolverSettings.UseHierarchy)
            //{
```

```
        //    naSolverSettings.HierarchyAttributeName="hierarchy";
        //    naSolverSettings.HierarchyLevelCount=3;
        //    naSolverSettings.set_MaxValueForHierarchy(1,1);
        //    naSolverSettings.set_NumTransitionToHierarchy(1,9);

        //    naSolverSettings.set_MaxValueForHierarchy(2,2);
        //    naSolverSettings.set_NumTransitionToHierarchy(2,9);
        //}
    }

    //更新分析上下文
    UpdateContextAfterSettingsChanged(NAContext);
}

//最近服务设施分析参数
public static void SetClosestFacilitySolverSettings(INAContext NACon-
text,string ImpedanceName,int targetCount)
{
    INASolver naSolver=NAContext.Solver;

    //Set specific Settings
    INAClosestFacilitySolver cfSolver=naSolver as INAClosestFacil-
itySolver;
    {
        ////停止遍历的默认截止值(The default cutoff value to stop traversing)
        //cfSolver.DefaultCutoff=1;
        //设置要查找的设施点的数目
        cfSolver.DefaultTargetFacilityCount=targetCount;
        //设置生成线的类型
        cfSolver.OutputLines=esriNAOutputLineType.esriNAOutputLi-
neTrueShapeWithMeasure;
        //设置遍历方向
        cfSolver.TravelDirection=esriNATravelDirection.esriNATravel-
DirectionToFacility;
    }

    //Generic Settings
```

```
        INASolverSettings naSolverSettings=naSolver as INASolverSettings;
        {
            //设置分析阻抗属性(Set the impedance attribute)
            naSolverSettings.ImpedanceAttributeName=ImpedanceName;

            ////设置单行线限制
            //IStringArray restrictions=naSolverSettings.Restriction-
AttributeNames;
            //restrictions.RemoveAll();
            //restrictions.Add("oneway");
            //naSolverSettings.RestrictionAttributeNames=restrictions;

            //设置允许 U 形转弯
            naSolverSettings.RestrictUTurns =
                                        esriNetworkForwardStarBack-
track.esriNFSBAllowBacktrack;
            //设置忽略无效位置
            naSolverSettings.IgnoreInvalidLocations=true;
        }

        //更新分析上下文
        UpdateContextAfterSettingsChanged(NAContext);
    }

    //服务范围分析参数设置
    public static void SetServiceAreaSolverSettings(INAContext NAContext,
string ImpedanceName,IDoubleArray BreaksArr)
    {
        //Set specific Settings
        INAServiceAreaSolver naSASolver=NAContext.Solver as INAServiceAr-
eaSolver;
        {
            //加载中断
            naSASolver.DefaultBreaks=BreaksArr;

            //设置生成多边形的类型
            naSASolver.OutputPolygons=esriNAOutputPolygonType.esriNAOut-
```

```
putPolygonSimplified;
            naSASolver.SplitPolygonsAtBreaks = false;
            naSASolver.MergeSimilarPolygonRanges = true;

            //设置生成线的类型
            //naSASolver.OutputLines = esriNAOutputLineType.esriNAOut-
putLineNone;
            naSASolver.OutputLines = esriNAOutputLineType.esriNAOutput-
LineTrueShape;
            naSASolver.OverlapLines = true;
            naSASolver.SplitLinesAtBreaks = false;

            //设置遍历方向
            naSASolver.TravelDirection = esriNATravelDirection.esriNATrav-
elDirectionFromFacility;
        }

        //Generic Settings
        INASolverSettings naSolverSettings = naSASolver as INASolverSet-
tings;
        {
            //设置分析阻抗属性(Set the impedance attribute)
            naSolverSettings.ImpedanceAttributeName = ImpedanceName;

            ////设置单行线限制
            //IStringArray restrictions = naSolverSettings.Restriction-
AttributeNames;
            //restrictions.RemoveAll();
            //restrictions.Add("Oneway");
            //naSolverSettings.RestrictionAttributeNames = restrictions;

            //设置禁止 U 形转弯
            naSolverSettings.RestrictUTurns = esriNetworkForwardStarBack-
track.esriNFSBNoBacktrack;
            //设置忽略无效位置
            naSolverSettings.IgnoreInvalidLocations = true;
        }
```

```
        //更新分析上下文
        UpdateContextAfterSettingsChanged(NAContext);
    }

    //更新分析上下文函数
    public static void UpdateContextAfterSettingsChanged(INAContext NA-
Context)
    {
        IDatasetComponent datasetComponent = NAContext.NetworkDataset
as IDatasetComponent;
        IDENetworkDataset deNetworkDataset = datasetComponent.DataEle-
ment as IDENetworkDataset;
        NAContext.Solver.UpdateContext(NAContext, deNetworkDataset,
new GPMessagesClass());
    }
```

(4)Solve()函数

本函数执行上下文的分析器 Solver 的 Solve()函数，返回过程信息。源代码如下：

```
    public static string Solve(INAContext pNAContext)
    {
        string errStr="";
        IGPMessages gpMessages=new GPMessagesClass();
        try
        {
            errStr="Error when solving";
            bool isPartialSolution=pNAContext.Solver.Solve(pNAContext,
gpMessages,null);

            if (!isPartialSolution)
                errStr="OK";
            else
                errStr="Partial Solution";
        }
        catch (Exception e)
        {
            errStr +=" Error Description "+e.Message;
        }
        return errStr;
    }
```

(5) getTemporaryPath()函数

本函数根据应用程序执行文件目录，返回临时数据库所在路径：

```
public static string getTemporaryPath(string path)
{
    int t;
    for (t=0; t < path.Length; t++)
    {
        if (path.Substring(t,4)= ="code")
        {
            break;
        }
    }
    string name=path.Substring(0,t - 1);
    return name;
}
```

15.4 功能调用

1. 添加类成员变量

在 MainForm 中添加类成员变量：

private INetworkDatasetm_ NetworkDataset=null;

stringpath=System. AppDomain. CurrentDomain. SetupInformation. ApplicationBase;

2. 添加启动页

在 RibbonControl 添加【Network】Tab 页，建立 Click 响应函数；此函数判断是否有打开的网络数据集，如果有，就赋值 m_ NetworkDataset 变量进行记录，并将 ribbonBarNetwork 设为可见。

源代码如下：

```
private void ribbonTabNetwork_Click(object sender,EventArgs e)
{
    //NetworkDataset RibbonBar
    ribbonBarNetwork.Enabled=StartingNetworkDatasetAnalyis();
    //UtilityNetwork RibbonBar
    ribbonBarUtilityNetwork.Enabled=StartingUtilityNetworkAnalysis();
}

private bool StartingNetworkDatasetAnalyis()
{
    IMap ipMap=_AxMapControl.ActiveView.FocusMap;
```

```
    if (ipMap.LayerCount==0)
    {
        MessageBox.Show("没有图层不能进行网络分析");
        return false;
    }
    for (int i=0; i < ipMap.LayerCount; i++)
    {
        ILayer ipLayer=ipMap.get_Layer(i);
        INetworkLayer ipFLayer=ipLayer as INetworkLayer;
        if (ipFLayer!=null)
        {
            m_NetworkDataset=ipFLayer.NetworkDataset;
            return true;
        }
    }
    return false;
}
```

3. 添加工具条消息响应函数

代码如下:

```
private void btnNetworkDatasetSolve_Click(object sender,EventArgs e)
{
    string NASelectedString="";
     foreach (DevComponents.DotNetBar.CheckBoxItem box in btnNet-
workMethod.SubItems)
    {
        if (box.Checked)
        {
            NASelectedString=box.Text;
            break;
        }
    }

    SolveTool pCommand;
    pCommand=new SolveTool(m_NetworkDataset,NASelectedString);
    pCommand.OnCreate(_AxMapControl.Object);
    pCommand.OnClick();
}
```

```
private void btnAddStops_Click(object sender,EventArgs e)
{
    ICommand pCommand;
    pCommand=new AddNetStopsTool();
    pCommand.OnCreate(_AxMapControl.Object);
    _AxMapControl.CurrentTool=pCommand as ITool;
    pCommand=null;
}

private void btnAddBarriers_Click(object sender,EventArgs e)
{
    ICommand pCommand;
    pCommand=new AddNetBarriesTool();
    pCommand.OnCreate(_AxMapControl.Object);
    _AxMapControl.CurrentTool=pCommand as ITool;
    pCommand=null;
}

private void btnAddFacility_Click(object sender,EventArgs e)
{
    ICommand pCommand;
    pCommand=new AddFacilitiesTool();
    pCommand.OnCreate(_AxMapControl.Object);
    _AxMapControl.CurrentTool=pCommand as ITool;
    pCommand=null;
}

private void btnAddIncidents_Click(object sender,EventArgs e)
{
    ICommand pCommand;
    pCommand=new AddIncidentsTool();
    pCommand.OnCreate(_AxMapControl.Object);
    _AxMapControl.CurrentTool=pCommand as ITool;
    pCommand=null;
}

private void btnNetworkClear_Click(object sender,EventArgs e)
{
```

```
        UtilityNetworkAnalyst.ClearElements(_AxMapControl.ActiveView,"
All");
        btnAddIncidents_Click(sender,e);
        btnAddFacility_Click(sender,e);
        btnAddBarriers_Click(sender,e);
        btnAddStops_Click(sender,e);
        _AxMapControl.CurrentTool=null;
    }
```

15.5　运行测试

按下 F5 键，编译运行程序。

第 16 章　几何网络分析

16.1　知识要点

本章介绍几何网络(Geometric Network)追踪分析等内容。网络结构数据直接在 ArcCatalog 中创建。

几何网路分析的思路分为以下步骤：

①打开 Geometric Network 网络；

②创建追踪器，相关接口 ITraceFlowSolverGEN；

③向追踪器添加节点标记元素(追踪器函数：PutJunctionOrigins(...))，边标记元素(PutEdgeOrigins(...))，添加节点/边障碍元素(set_ElementBarriers(...))；

④执行追踪计算，使用追踪器函数，如 FindFlowElements()，FindPath()等；

⑤显示结果信息。

16.2　功能描述

点击【Network】Tab 页。如果几何网络已经加载到视图(可用 Add Data 工具加载)，【Utility Network Analysis】RibbonBar 上所有按钮有效，如图 16-1 所示。

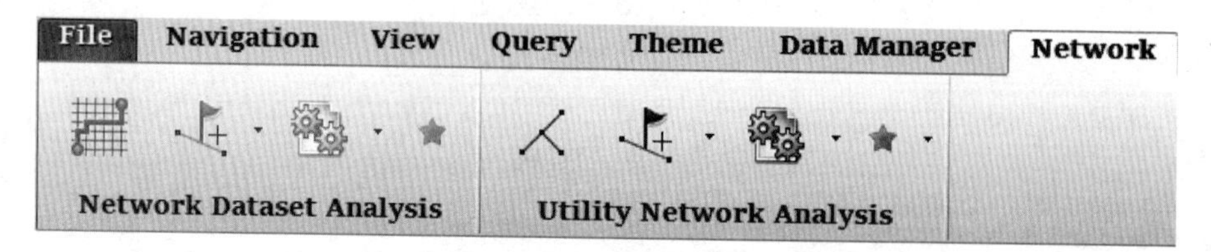

图 16-1　Ribbon Bar 上的按钮

包括 4 个按钮(从左到右顺序)，作用如下：

①求解(btnUtilityNASolve)按钮：分析求解。

②位置添加下拉菜单(ButtonItem 类型)，点击右侧下三角，如图 16-2(a)所示：

◆　添加节点标志(Add JunctionFlag)；

◆　添加边标志(Add EdgeFlag)；

◆　添加节点障碍(Add Junction Barriers)；

◆　添加边障碍(Add Edge Barriers)。

③方法选择下拉菜单(CheckBoxItem 类型)，如图 16-2(b)所示。

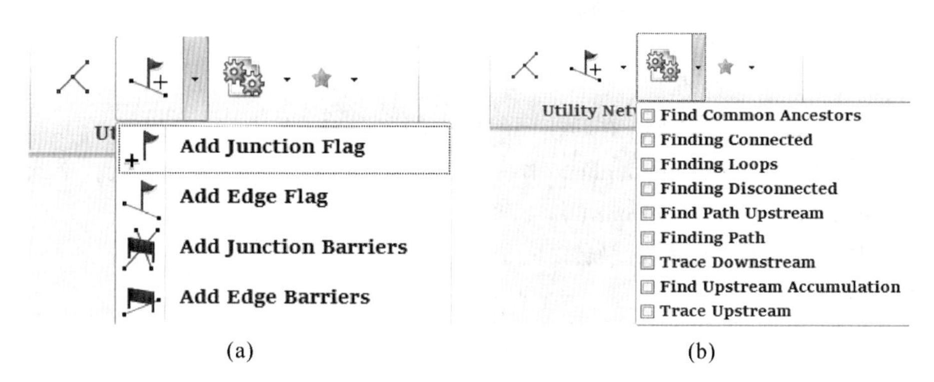

(a)　　　　(b)

图 16-2　下拉菜单

可选择方法及含义见表 16-1。

表 16-1　**可选择方法及含义**

方法名	对应追踪器方法	注　　释
Find Common Ancestors	FindCommonAncestors	查找共同祖先
Finding Connected	FindFlowElements	查找相连接的网络要素
Finding Loops	FindCircuits	查找网络中的环
Finding Disconnected	FindFlowUnreachedElements	查找未连接的网络要素
Find Path Upstream	FindSource	查找上游路径
Finding Path	FindPath	查找路径
Trace Downstream	FindFlowElements	下游追踪
Find Upstream Accumulation	FindAccumulation	查找上游路径累积消耗
Trace Upstream	FindSource	上游追踪

④清除下拉菜单(ButtonItem 类型)：Clear Flag/Clear Barries/Clear Result，分别清除标志点、障碍点、分析结果。

16.3　功能实现

16.3.1　工具条功能实现

1. 节点标志/边标志添加功能

节点标志/边标志添加功能激活后，光标在屏幕上拾取点，捕捉与之最近的节点/最近的边及其相应的位置点信息，如果在容差范围内没有捕捉到网络元素，则操作被忽略。

之后，在位置点上创建一个 Mark 类型的图形元素，显示节点标志/边标志符号；同时将其转化 Point 要素对象分别存入临时数据库(TemporaryGeodatabase. gdb)的 JunctionFlag、EdgeFlag 要素类中，要素类中 EID 属性字段记录捕捉到的网络元素的 FID，供之后分析用。

功能类继承 BaseTool 实现，在新建项的模板浏览窗口，选择 ArcEngine 的 BaseTool 模板，功能类取名为 ToolAddJunctionFlag/ToolAddEdgeFlag。新工具框架生成后，重载鼠标响应函数：OnClick()、OnMouseDown()。

ToolAddJunctionFlag 关键源代码如下：

```
//定义私有成员
private IHookHelper m_hookHelper;
private IFeatureClass flagFClass;
string path = System.AppDomain.CurrentDomain.SetupInformation.ApplicationBase;
private IGeometricNetwork m_geometricNetwork;
//定义几何网络属性(只写)
public IGeometricNetwork GeometricNetwork
{
    set
    {
        m_geometricNetwork=value;
    }
}

//重载点击按钮事件函数:打开 TemporaryGeodatabase.gdb 中的 JunctionF 要素类,并清空
public override void OnClick()
{
    // TODO:Add ToolAddJunctionFlag.OnClick implementation
    string strName =NetworkAnalyst.getTemporaryPath (path) +
                                                    " \\data \\HuanbaoGeodatabase.gdb";
    IFeatureWorkspace pFWorkspace =NetworkAnalyst.OpenWorkspace(strName)
                                                    as IFeatureWorkspace;
    flagFClass = pFWorkspace.OpenFeatureClass (LocationTypeEnum.JunctionFlag.ToString());
    if (flagFClass.FeatureCount(null) > 0)
```

```
        {
            ITable pTable=flagFClass as ITable;
            pTable.DeleteSearchedRows(null);
        }
    }
    //重载鼠标点击事件函数
    public override void OnMouseDown(int Button,int Shift,int X,int Y)
    {
        //获取鼠标点击位置的空间点
        IActiveView activeView=m_hookHelper.ActiveView;
        IPoint point =activeView.ScreenDisplay.DisplayTransformation.
ToMapPoint(X,Y);

        //IPointToEID接口用来通过空间点获取该位置的几何网络元素
        IPointToEID pointToEID=new PointToEIDClass();
        pointToEID.GeometricNetwork=m_geometricNetwork;
        pointToEID.SourceMap=m_hookHelper.FocusMap;
        pointToEID.SnapTolerance=10;

        //通过IPointToEID接口的GetNearestJunction方法获取点击位置最邻近的管点
元素
        int nearestJunctionEID=-1;
        IPoint outPoint=new PointClass();
        pointToEID.GetNearestJunction(point,out nearestJunctionEID,out out-
Point);

        //如果获取得到管点元素
        if (outPoint!=null)
        {
            //创建一个点要素,记录障碍位置
            IFeature pFeature=flagFClass.CreateFeature();
            {
                pFeature.Shape=point;
                pFeature.set_Value(2,nearestJunctionEID);
                pFeature.Store();
            }

            //绘制并显示管点标识
```

```
            IElement element=new MarkerElementClass();
            {
                ISimpleMarkerSymbol simpleMarkerSymbol=new SimpleMark-
erSymbolClass();
                simpleMarkerSymbol.Color=ColorToIRgbColor(Color.Fro-
mArgb(85,255,0));
                simpleMarkerSymbol.Size=12;
                simpleMarkerSymbol.Style = esriSimpleMarkerStyle.esriSMS-
Diamond;

                ((IMarkerElement)element).Symbol=simpleMarkerSymbol;
                ((IElementProperties)element).Name="Flag";
                element.Geometry=outPoint;
            }

            ((IGraphicsContainer)activeView).AddElement(element,0);
            activeView.PartialRefresh(esriViewDrawPhase.esriViewGraphics,
null,activeView.Extent);
        }
    }
```

ToolAddEdgeFlag 关键源代码如下:

```
    //定义私有成员
    private IHookHelper m_hookHelper;
    private IFeatureClass flagFClass;
    string path=System.AppDomain.CurrentDomain.SetupInformation.Ap-
plicationBase;
    private IGeometricNetwork m_geometricNetwork;
    //定义几何网络属性(只写)
    public IGeometricNetwork GeometricNetwork
    {
        set
        {
            m_geometricNetwork=value;
        }
    }

    //重载点击按钮事件函数:打开 TemporaryGeodatabase.gdb 中的 JunctionF 要素类,并
清空
```

```
    public override void OnClick()
    {
        // TODO:Add ToolAddEdgeFlag.OnClick implementation
        string strName=NetworkAnalyst.getTemporaryPath(path ) +" \\data
\\HuanbaoGeodatabase.gdb";
        IFeatureWorkspace pFWorkspace=NetworkAnalyst.OpenWorkspace(strName)
                                                as IFeatureWorkspace;
        flagFClass=pFWorkspace.OpenFeatureClass(LocationTypeEnum.Edge-
Flag.ToString());
        if (flagFClass.FeatureCount(null) > 0)
        {
            ITable pTable=flagFClass as ITable;
            pTable.DeleteSearchedRows(null);
        }
    }

    //重载鼠标点击事件函数
    public override void OnMouseDown(int Button,int Shift,int X,int Y)
    {
        //获取鼠标点击位置的空间点
        IActiveView activeView=m_hookHelper.ActiveView;
        IPoint point=activeView.ScreenDisplay.DisplayTransformation.
ToMapPoint(X,Y);

        //IPointToEID 接口用来通过空间点获取该位置的几何网络元素
        IPointToEID pointToEID=new PointToEIDClass();
        pointToEID.GeometricNetwork=m_geometricNetwork;
        pointToEID.SourceMap=m_hookHelper.FocusMap;
        pointToEID.SnapTolerance=10;

        //通过 IPointToEID 接口的 GetNearestJunction 方法获取点击位置最邻近的管线
元素
        int nearestEdgeEID=-1;
        IPoint outPoint=new PointClass();
        double percent=-1;
        pointToEID.GetNearestEdge(point,out nearestEdgeEID,out outPoint,
out percent);
```

```
        //如果获取得到管线元素
        if (outPoint!=null)
        {
            //创建一个点要素,记录障碍位置;
            IFeature pFeature=flagFClass.CreateFeature();
            {
                pFeature.Shape=point;
                pFeature.set_Value(2,nearestEdgeEID);
                pFeature.Store();
            }

            //绘制并显示管线标识
            IElement element=new MarkerElementClass();
            {
                 ISimpleMarkerSymbol simpleMarkerSymbol=new SimpleMark-
erSymbolClass();
                simpleMarkerSymbol.Color=ColorToIRgbColor(Color.FromArgb
(85,255,0));
                simpleMarkerSymbol.Size=12;
                simpleMarkerSymbol.Style=esriSimpleMarkerStyle.esriS-
MSSquare;

                ((IMarkerElement)element).Symbol=simpleMarkerSymbol;
                ((IElementProperties)element).Name="Flag";
                element.Geometry=outPoint as IGeometry;
            }
            ((IGraphicsContainer)activeView).AddElement(element,0);
            activeView.PartialRefresh(esriViewDrawPhase.esriViewGraphics,
null,activeView.Extent);
        }
    }
```

注意：节点标注点、边标志的添加功能，主要区别在于寻找最近网络元素使用不同方法，前者使用 IPointToEID 的 GetNearestJunction(...)寻找最近 Junction，后者使用 GetNearestEdge(...)寻找最近 Edge。

节点事故点/边事故点的添加按钮功能，完全类似于节点标志点/边标志点添加功能，不同的是存入到临时数据库的不同的数据集 JunctionBarries、EdgeBarries 中。

2. 分析(ToolSolve)功能

ToolSolve 功能类继承 BaseCommand 实现，在新建项的模板浏览窗口，选择 ArcEngine

的 Base Command 模板，功能类取名为 ToolSolve。新工具框架生成后，重载鼠标响应函数 OnClick()；本函数根据 m_NASelectedIndex 变量值记录不同的追踪类型。

```
//添加类成员变量
private IHookHelper m_hookHelper=null;
private IGeometricNetwork m_geometricNetwork=null;
private int m_NASelectedIndex=0;
string path = System.AppDomain.CurrentDomain.SetupInformation.Ap-
plicationBase;

//修改构造函数
public ToolSolve(IGeometricNetwork geometricNetwork,int index)
{
    m_geometricNetwork=geometricNetwork;
    m_NASelectedIndex=index;

    base.m_category="NetWorkAnalyst"; //localizable text
    base.m_caption="Solve";  //localizable text
    base.m_message="Solve";  //localizable text
    base.m_toolTip="Solve";  //localizable text
    base.m_name="SolveTool";   //unique id,non-localizable (e.g. "
MyCategory_MyCommand")

    try
    {
        string bitmapResourceName=GetType().Name+".bmp";
        base.m_bitmap=new Bitmap(GetType(),bitmapResourceName);
    }
    catch (Exception ex)
    {
         System.Diagnostics.Trace.WriteLine(ex.Message,"Invalid
Bitmap");
    }
}
```

重载 OnClick()函数，实现分析功能，步骤如下：

①打开节点标志/边标志/节点障碍点/边障碍点要素类；

②创建追踪器；

③添加节点标志/边标志/节点障碍点/边障碍点元素；

④执行追踪计算；

⑤显示分析结果。

OnClick( )源代码以及相关辅助函数如下(这里要用到 UtilityNetworkAnalyst, NetworkAnalyst 核心功能类, 后文将介绍):

```
public override void OnClick()
{
    // TODO:Add SolveTool.OnClick implementation
    string strName=NetworkAnalyst.getTemporaryPath(path) +"\\data\\HuanbaoGeodatabase.gdb";
    IFeatureWorkspace pFWorkspace = NetworkAnalyst.OpenWorkspace(strName) as IFeatureWorkspace;
    IFeatureClass JunctionFClass=
                pFWorkspace.OpenFeatureClass(LocationTypeEnum.JunctionFlag.ToString());
    IFeatureClass EdgeFClass=
                pFWorkspace.OpenFeatureClass(LocationTypeEnum.EdgeFlag.ToString());
    IFeatureClass JunctionBarriesFClass=
                pFWorkspace.OpenFeatureClass(LocationTypeEnum.JunctionBarriers.ToString());
    IFeatureClass EdgeBarriesFClass=
                pFWorkspace.OpenFeatureClass(LocationTypeEnum.EdgeBarriers.ToString());

    //创建追踪器
     ITraceFlowSolverGEN traceFlowSolverGEN = new TraceFlowSolverClass();
    INetSolver netSolver=traceFlowSolverGEN as INetSolver;
    netSolver.SourceNetwork=m_geometricNetwork.Network;

    UtilityNetworkAnalyst.GeometricNetwork=m_geometricNetwork;
    //添加节点元素
    IJunctionFlag[] JunctionFlagArr=
                            UtilityNetworkAnalyst.LoadJunctionFlagLocation(JunctionFClass);
    int junctionCount=JunctionFlagArr.Length;
    traceFlowSolverGEN.PutJunctionOrigins(ref JunctionFlagArr);

    //添加边元素
```

```
        IEdgeFlag[] EdgeFlagArr=UtilityNetworkAnalyst.LoadEdgeFlagLo-
cation(EdgeFClass);
        int edgeCount=EdgeFlagArr.Length;
        traceFlowSolverGEN.PutEdgeOrigins(ref EdgeFlagArr);

        //添加节点/边障碍元素
        INetElementBarriers netBarriersGEN=null;
        netBarriersGEN=UtilityNetworkAnalyst.LoadJunctionBarriers(Junc-
tionBarriesFClass)
                                                  as INetElementBarriers;
        netSolver.set_ElementBarriers(esriElementType.esriETJunction,
netBarriersGEN);
        netBarriersGEN=UtilityNetworkAnalyst.LoadEdgeBarriers(Edge-
BarriesFClass)
                                                  as INetElementBarriers;
        netSolver.set_ElementBarriers(esriElementType.esriETEdge,net-
BarriersGEN);

        //执行追踪计算
        IEnumNetEID junctionEIDs=null;
        IEnumNetEID edgeEIDs=null;
        string sCost=UtilityNetworkAnalyst.TraceFlowSolver(traceFlow-
SolverGEN,m_NASelectedIndex,
                                         junctionCount,edgeCount,out
junctionEIDs,out edgeEIDs);

        //首先清除已有的分析结果
        UtilityNetworkAnalyst.ClearElements(m_hookHelper.ActiveView,"Re-
sult");

        //结果化图形元素
        IColor pColor=ColorToIRgbColor(Color.FromArgb(255,0,0));
        if(junctionEIDs.Count >0 )
            GeographicJunctionResults(junctionEIDs,pColor);
        if (edgeEIDs.Count > 0)
            GeographicEdgeResults(edgeEIDs,pColor);

        //刷新地图中的图形
```

```
        IActiveView pActiveView=m_hookHelper.ActiveView;
        pActiveView. PartialRefresh ( esriViewDrawPhase. esriViewGraphics,
null,pActiveView.Extent);
    }
    //根据设定的颜色绘制网络分析结果中的管点
    private void GeographicJunctionResults ( IEnumNetEID junctionEIDs,
IColor color)
    {
        IActiveView pActiveView=m_hookHelper.ActiveView;
        //设置管点显示的 Symbol
        ISimpleMarkerSymbol simpleMarkerSymbol=new SimpleMarkerSymbolClass
();
        {
            simpleMarkerSymbol.Color=color;
            simpleMarkerSymbol.Size=6;
            simpleMarkerSymbol.Style=esriSimpleMarkerStyle.esriSMSCircle;
        }
        //创建 EID 帮助对象
        IEIDHelper pEIDHelper=new EIDHelperClass();
        {
            pEIDHelper.GeometricNetwork=m_geometricNetwork;
            //pEIDHelper.OutputSpatialReference=pSpatialReference;
            pEIDHelper.ReturnGeometries=true;
        }

        //根据节点的 ID 获取边的信息
        IEnumEIDInfo pEnumEIDInfo=pEIDHelper.CreateEnumEIDInfo(junc-
tionEIDs);
        pEnumEIDInfo.Reset();
        IEIDInfo pEIDInfo;
        while ((pEIDInfo=pEnumEIDInfo.Next())!=null)
        {
            IElement element=new MarkerElementClass();
            element.Geometry=pEIDInfo.Geometry;
            ((IMarkerElement)element).Symbol=simpleMarkerSymbol;
            ((IElementProperties)element).Name="Result";
            ((IGraphicsContainer)pActiveView).AddElement(element,0);
        }
```

```
    }
    //根据设定的颜色绘制网络分析结果中的管线
    private void GeographicEdgeResults ( IEnumNetEID edgeEIDs, IColor
color)
    {
        IActiveView pActiveView=m_hookHelper.ActiveView;

        //设置管点显示的 Symbol
        ISimpleLineSymbol simpleLineSymbol=new SimpleLineSymbolClass();
        {
            simpleLineSymbol.Color=color;
            simpleLineSymbol.Width=2;
            simpleLineSymbol.Style=esriSimpleLineStyle.esriSLSSolid;
        }

        //创建 EID 帮助对象
        IEIDHelper pEIDHelper=new EIDHelperClass();
        {
            pEIDHelper.GeometricNetwork=m_geometricNetwork;
            //pEIDHelper.OutputSpatialReference=pSpatialReference;
            pEIDHelper.ReturnGeometries=true;
        }

        //根据边的 ID 获取边的信息
        IEnumEIDInfo pEnumEIDInfo=pEIDHelper.CreateEnumEIDInfo(edgeE-
IDs);
        pEnumEIDInfo.Reset();
        IEIDInfo pEIDInfo;
        while ((pEIDInfo=pEnumEIDInfo.Next())!=null)
        {
            IElement element=new LineElementClass();
            element.Geometry=pEIDInfo.Geometry;
            ((ILineElement)element).Symbol=simpleLineSymbol;
            ((IElementProperties)element).Name="Result";
            ((IGraphicsContainer)pActiveView).AddElement(element,0);
        }
    }
```

16.3.2 核心功能类的实现

为方便操作，将网络分析的核心功能封装为 UtilityNetWorkAnalyst，共有成员为静态成员，方便直接调用。

1. UtilityNetWorkAnalyst 设计

代码如下：

```
public class UtilityNetworkAnalyst
{
    private static IGeometricNetwork m_geometricNetwork=null;
    public static IGeometricNetwork GeometricNetwork
    {
        set { m_geometricNetwork=value; }
    }

    //加载节点标志信息
     public static IJunctionFlag[ ] LoadJunctionFlagLocation(IFea-
tureClass inputFC)

    //加载边标志信息
     public static IEdgeFlag[ ] LoadEdgeFlagLocation(IFeatureClass
inputFC)

    //加载网络障碍信息
     public static INetElementBarriers LoadNetworkBarriers(IFea-
tureClass inputFC,esriElementTypeelementType)

    //追踪计算
    public static string TraceFlowSolver(ITraceFlowSolverGEN trace-
FlowSolverGEN,
                              int SelectedIndex, int JunctionFlagsCount,
int EdgeFlagsCount,
                              out IEnumNetEID junctionEIDs,out IEnumNetEID
edgeEIDs)
    //若干辅助函数
    //获取临时数据库存放路径
    public static string getTemporaryPath(string path)
    //清除视图中的图形元素
    public static void ClearElements(IActiveView activeView,string
```

```
elementName)
        ……
    }
```

位置枚举类型 LocationTypeEnum 扩充如下(注意要求临时数据库中相应的要素类名称与枚举名称要相同):

```
    public enum LocationTypeEnum
    {
        JunctionFlag=6,
        EdgeFlag=7,
        JunctionBarriers=8,
        EdgeBarriers=9,
        BurstPipe=10,
    }
```

2. UtilityNetWorkAnalyst 实现

(1)LoadJunctionFlagLocation()函数

函数根据已经打开的节点标志要素类,获取节点标志结构数组 IJunctionFlag[],源代码如下:

```
    public static IJunctionFlag[] LoadJunctionFlagLocation(IFeature-
Class inputFC)
    {
        //获取 INetElements 接口
        INetElements netElements=m_geometricNetwork.Network as INetEle-
ments;
        int userClassID=0;
        int userID=0;
        int userSubID=0;

        //初始化管点标志数组
        int count=inputFC.FeatureCount(null);
        IJunctionFlag[] JunctionFlagArr=new IJunctionFlag[count];

        //遍历节点标志要素类,将其转化为管点标志数组
        IFeatureCursor pFCursor=inputFC.Search(null,false);
        IFeature pFeature=null;
        int i=0;
        while ((pFeature=pFCursor.NextFeature())!=null)
        {
            int nearestJunctionEID=(int)pFeature.get_Value(2);
```

```
            netElements.QueryIDs(nearestJunctionEID, esriElementType.
esriETJunction,
                                                              out userClassID,
out userID,out userSubID);
            //创建管点标识,并将其加入管点标识数组中
            INetFlag junctionFlag=new JunctionFlagClass() as INetFlag;
            junctionFlag.UserClassID=userClassID;
            junctionFlag.UserID=userID;
            JunctionFlagArr[i]=junctionFlag as IJunctionFlag;
            i++;
        }

        return JunctionFlagArr;
    }
```

(2)LoadEdgeFlagLocation()函数

函数根据已经打开的边标志要素类，获取边标志结构数组 IEdgeFlag[]，源代码如下：

```
    public static IEdgeFlag[] LoadEdgeFlagLocation(IFeatureClass inputFC)
    {
        //获取 INetElements 接口
        INetElements netElements=m_geometricNetwork.Network as INetEle-
ments;
        int userClassID=0;
        int userID=0;
        int userSubID=0;

        //初始化边标志数组
        int count=inputFC.FeatureCount(null);
        IEdgeFlag[] EdgeFlagArr=new IEdgeFlag[count];

        //遍历边标志要素类,将其转化为边标志数组
        IFeatureCursor pFCursor=inputFC.Search(null,false);
        IFeature pFeature=null;
        int i=0;
        while ((pFeature=pFCursor.NextFeature())!=null)
        {
            int nearestEdgeEID=(int)pFeature.get_Value(2);
            netElements.QueryIDs(nearestEdgeEID,esriElementType.esri-
ETEdge,out userClassID,out userID,out userSubID);
```

```
            //创建管线标识,并将其加入管线标识数组中
            INetFlag edgeFlag=new EdgeFlagClass() as INetFlag;
            edgeFlag.UserClassID=userClassID;
            edgeFlag.UserID=userID;
            EdgeFlagArr[i]=edgeFlag as IEdgeFlag;
            i++;
        }

        return EdgeFlagArr;
    }
```

(3) LoadNetworkBarriers()函数

代码如下:

```
    public static INetElementBarriers LoadNetworkBarriers(IFeatureClass
inputFC,esriElementType elementType)
    {
        int count=inputFC.FeatureCount(null);
        //如果目前有管点障碍,则加入分析器中
        if (count > 0)
        {
            int[] BarrierEIDs=new int[count];

            //遍历障碍要素类,将EID属性转化为整型数组
            IFeatureCursor pFCursor=inputFC.Search(null,false);
            IFeature pFeature=null;
            int i=0;
            while ((pFeature=pFCursor.NextFeature())!=null)
            {
                BarrierEIDs[i]=(int)pFeature.get_Value(2);
                i++;
            }

            //创建网络障碍接口对象
            INetElementBarriersGEN netElementBarriersGEN=new NetElement-
BarriersClass();
            netElementBarriersGEN.Network=m_geometricNetwork.Network;
            netElementBarriersGEN.ElementType=elementType;
            netElementBarriersGEN.SetBarriersByEID(ref BarrierEIDs);
```

```
            return netElementBarriersGEN as INetElementBarriers;
        }
        //否则将障碍设置为空
        else
        {
            return null;
        }
    }
```

(4) TraceFlowSolver()函数

本函数根据方法索引号(SelectedIndex)，执行追踪器 ITraceFlowSolverGEN 的对应函数，返回累积耗费数据。源代码如下：

```
    public static string TraceFlowSolver ( ITraceFlowSolverGEN trace-
FlowSolverGEN,int SelectedIndex,int JunctionFlagsCount,int EdgeFlag-
sCount,out IEnumNetEID junctionEIDs,out IEnumNetEID edgeEIDs)
    {
        //定义 EnumNetEIDArrayClass 变量,用于记录追踪路线经过的管点和管边
        junctionEIDs=new EnumNetEIDArrayClass();
        edgeEIDs=new EnumNetEIDArrayClass();
        string strCosts="";
        try
        {
            int count=-1;
            object[] segmentCosts=null;
            object pTotalCost=null;
            switch (SelectedIndex)
            {
                case 0://查找共同祖先
                    traceFlowSolverGEN.FindCommonAncestors(esriFlow-
Elements.esriFEJunctionsAndEdges,out junctionEIDs,out edgeEIDs);
                    break;
                case 1://查找相连接的网络要素
                    traceFlowSolverGEN.FindFlowElements(esriFlowMethod.
esriFMConnected,esriFlowElements.esriFEJunctionsAndEdges,out junctio-
nEIDs,out edgeEIDs);
                    break;
                case 2://查找网络中的环
```

```
                    traceFlowSolverGEN.FindCircuits (esriFlowElements.
esriFEJunctionsAndEdges,out junctionEIDs,out edgeEIDs);
                    break;
                case 3://查找未连接的网络要素
                    traceFlowSolverGEN.FindFlowUnreachedElements(esri-
FlowMethod.esriFMConnected,esriFlowElements.esriFEJunctionsAndEdges,
out junctionEIDs,out edgeEIDs);
                    break;
                case 4://查找上游路径,同时获取网络追踪的耗费
                    count =JunctionFlagsCount+EdgeFlagsCount;
                    if (count > 0)
                    {
                        segmentCosts=new object[count];
                        traceFlowSolverGEN.FindSource(esriFlowMethod.
esriFMUpstream, esriShortestPathObjFn. esriSPObjFnMinSum, out junctio-
nEIDs,out edgeEIDs,count,ref segmentCosts);
                        strCosts=GetSegmentCosts(segmentCosts).ToString();
                    }
                    break;
                case 5://查找路径,同时获取网络追踪的耗费
                    //当同时存在 JunctionFlag 和 EdgeFlag 时,该功能不可用
                    if (JunctionFlagsCount > 0 && EdgeFlagsCount > 0)
                        break;
                    //count 比所有标识的总数少 1 个
                    count =Math.Max(JunctionFlagsCount,EdgeFlagsCount) -1;
                    if (count > 0)
                    {
                        segmentCosts=new object[count];
                        traceFlowSolverGEN.FindPath(esriFlowMethod.es-
riFMConnected, esriShortestPathObjFn. esriSPObjFnMinSum, out junctio-
nEIDs,out edgeEIDs,count,ref segmentCosts);
                        strCosts=GetSegmentCosts(segmentCosts).ToString();
                    }
                    break;
                case 6://下游追踪
                    traceFlowSolverGEN.FindFlowElem ents (esriFlowMethod.
esriFMDownstream, esriFlowElements. esriFEJunctionsAndEdges, out junc-
tionEIDs,out edgeEIDs);
```

```
                    break;
                case 7://查找上游路径累积消耗,同时获取网络追踪的耗费
                    pTotalCost=new object();
                    traceFlowSolverGEN.FindAccumulation(esriFlowMethod.es-
riFMUpstream,esriFlowElements.esriFEJunctionsAndEdges,out junctionEIDs,
out edgeEIDs,out pTotalCost);strCosts=pTotalCost.ToString();
                    break;
                case 8://上游追踪
                    count=JunctionFlagsCount+EdgeFlagsCount;
                    if (count > 0)
                    {
                        segmentCosts=new object[count];
                        traceFlowSolverGEN.FindSource(esriFlowMethod.esriF-
MUpstream, esriShortestPathObjFn. esriSPObjFnMinSum, out junctionEIDs,
out edgeEIDs,count,ref segmentCosts);
                        strCosts="";
                    }
                    break;
            }
        }
        catch { }

        return strCosts;
    }
    //耗费汇总函数
    private static int GetSegmentCosts(object[] segmentCosts)
    {
        int sum=0;
        for (int i=0; i < segmentCosts.Length; i++)
            sum +=Convert.ToInt32(segmentCosts[i]);
        return sum;
    }
```

(5)ClearElements()函数

根据 Element 的名称清除当前视图中的 Element。代码如下:

```
    public static void ClearElements(IActiveView activeView,string ele-
mentName)
    {
        IGraphicsContainer graphicsContainer=activeView as IGraphics-
```

```
Container;
        graphicsContainer.Reset();
        IElement element=graphicsContainer.Next();
        while (element!=null)
        {
            if (elementName=="All")
                graphicsContainer.DeleteElement(element);

            if (((IElementProperties)element).Name==elementName)
                graphicsContainer.DeleteElement(element);

            element=graphicsContainer.Next();
        }
        activeView.Refresh();
    }
```

16.4　功能调用

1. 添加类成员变量

在 MainForm 中添加类成员变量:

private IGeometricNetworkm_geometricNetwork=null;

2. 添加启动菜单

在 RibbonControl 的【Network】Tab 页建立 Click 响应函数；此函数判断当前视图中是否有已打开的几何网络，如果有，就第一个网络数据结构赋值 m_geometricNetwork 变量，并将分析工具条 ribbonBarUtilityNetwork 设为有效。

源代码如下：

```
private void ribbonTabNetwork_Click(object sender,EventArgs e)
{
    //NetworkDataset RibbonBar
    ribbonBarNetwork.Enabled=StartingNetworkDatasetAnalyis();
    //UtilityNetwork RibbonBar
    ribbonBarUtilityNetwork.Enabled=StartingUtilityNetworkAnalysis();
}

private bool StartingUtilityNetworkAnalysis()
{
    IMap ipMap=_AxMapControl.ActiveView.FocusMap;
    if (ipMap.LayerCount==0)
```

```
        {
            MessageBox.Show("没有图层不能进行网络分析");
            return false;
        }

        for (int i=0; i < ipMap.LayerCount; i++)
        {
            IFeatureLayer ipFLayer=ipMap.get_Layer(i) as IFeatureLayer;
            if (ipFLayer!=null)
            {
                //获取 network 集合
                IFeatureDataset pFeatureDataset =ipFLayer.FeatureClass.Fea-
tureDataset;
                INetworkCollection pNetWorkCollection = pFeatureDataset
as INetworkCollection;

                //获取第一个 Geometric Network 对象
                int intNetworkCount =pNetWorkCollection.GeometricNetwork-
Count;
                if (intNetworkCount > 0)
                {
                    m_geometricNetwork=pNetWorkCollection.get_Geomet-
ricNetwork(0);
                    return true;
                }
            }
        }
        return false;
    }
```

3. 添加工具条消息响应函数

代码如下：

```
    private void btnUtilityNASolve_Click(object sender,EventArgs e)
    {
        int NASelectedIndex=0;
        foreach (DevComponents.DotNetBar.CheckBoxItem box in btnUtili-
tyNAMethod.SubItems)
        {
            if (box.Checked)  break;
```

```
            NASelectedIndex++;
        }

        ToolSolve toolSolve = new ToolSolve(m_geometricNetwork, NASe-
lectedIndex);
        toolSolve.OnCreate(_AxMapControl.Object);
        toolSolve.OnClick();
    }

    private void btnAddJunctionFlag_Click(object sender,EventArgs e)
    {
        ToolAddJunctionFlag toolAddJunctionFlag = new ToolAddJunctionF-
lag();
        toolAddJunctionFlag.GeometricNetwork=m_geometricNetwork;
        toolAddJunctionFlag.OnCreate(_AxMapControl.Object);
        _AxMapControl.CurrentTool=toolAddJunctionFlag as ITool;
    }
    private void btnAddEdgeFlag_Click(object sender,EventArgs e)
    {
        ToolAddEdgeFlag toolAddEdgeFlag=new ToolAddEdgeFlag();
        toolAddEdgeFlag.GeometricNetwork=m_geometricNetwork;
        toolAddEdgeFlag.OnCreate(_AxMapControl.Object);
        _AxMapControl.CurrentTool=toolAddEdgeFlag as ITool;
    }

    private void btnAddJunctionBarriers_Click(object sender,EventArgs e)
    {
        ToolAddJunctionBarriers toolAddJunctionBarriers=new ToolAddJunc-
tionBarriers();
        toolAddJunctionBarriers.GeometricNetwork=m_geometricNetwork;
        toolAddJunctionBarriers.OnCreate(_AxMapControl.Object);
        _AxMapControl.CurrentTool=toolAddJunctionBarriers as ITool;
    }

    private void btnAddEdgeBarriers_Click(object sender,EventArgs e)
    {
        ToolAddEdgeBarriers toolAddEdgeBarriers = new ToolAddEdgeBarri-
ers();
```

```
        toolAddEdgeBarriers.GeometricNetwork=m_geometricNetwork;
        toolAddEdgeBarriers.OnCreate(_AxMapControl.Object);
        _AxMapControl.CurrentTool=toolAddEdgeBarriers as ITool;
    }

    private void btnClearFlag_Click(object sender,EventArgs e)
    {
        UtilityNetworkAnalyst. ClearElements ( _ AxMapControl. ActiveView,
"Flag");
        btnAddEdgeFlag_Click(sender,e);
        btnAddJunctionFlag_Click(sender,e);
        _AxMapControl.CurrentTool=null;
    }

    private void btnClearBarriers_Click(object sender,EventArgs e)
    {
        UtilityNetworkAnalyst. ClearElements ( _ AxMapControl. ActiveView,
"Barriers");
        btnAddEdgeBarriers_Click(sender,e);
        btnAddJunctionBarriers_Click(sender,e);
        _AxMapControl.CurrentTool=null;
    }

    private void btnClearResult_Click(object sender,EventArgs e)
    {
        //清除各类标识显示图形,并将管点和管线标识列表清空
        UtilityNetworkAnalyst. ClearElements ( _ AxMapControl. ActiveView,
"Result");
    }
```

16.5 运行测试

按下 F5 键，编译运行程序。

第 17 章　属性数据表的查询显示

17.1　功能描述

从关系数据库的角度而言，ITable 对象代表了 GeoDatabase 中的一张二维表(或者视图 View)，一个表由多个列(IFields)定义，列被称为字段(IField)，IFields 是 IField 的集合。使用 ITable 对象 AddField(添加字段)、DeleteField(删除字段)方法，可改变表的结构。存储在表中的元素是 IRow 对象，一个 IRow 对象代表了表中的一条记录。

IFeatureClass 对象继承 ITable，所以它也是一个表，只是存在一个表示要素空间图形的特殊字段(Geometry)，一行记录代表一个要素。对于离散型栅格数据(IRaster)，每个波段存在(或可重建)表示“值-频数”统计信息的属性表(AttributeTable)，它属于 ITable 类型，因此，我们也可以用表格的形式打开一个要素类或打开栅格数据的属性表。

VS 中 DataGridView 控件提供了一种强大而灵活的以表格形式显示数据的方式。可以使用 DataGridView 控件来显示少量数据的只读视图，也可以对其进行缩放以显示特大数据集的可编辑视图。还可以很方便地把一个 DataTable 绑定到 DataGridView 控件数据源。利用 DataGridView 显示 ITable 数据，比较有效的办法是：

①先将 ITable 转换为 DataTable，包括创建 DataTable 结构、填充表体等；

②然后再将 DataTable 绑定到 DataGridView 控件数据源；

③最后调用并显示属性表窗体。

17.2　功能描述

在 ArcMap 中，单击图层右键菜单中的“Open Attribute Table”命令，便可弹出属性数据表。本章将完成类似的功能，效果如图 17-1 所示。

17.3　功能实现

1. 创建属性表窗体

新建一个 Windows 窗体，命名为“OpenAttributeTableFrm.cs”。

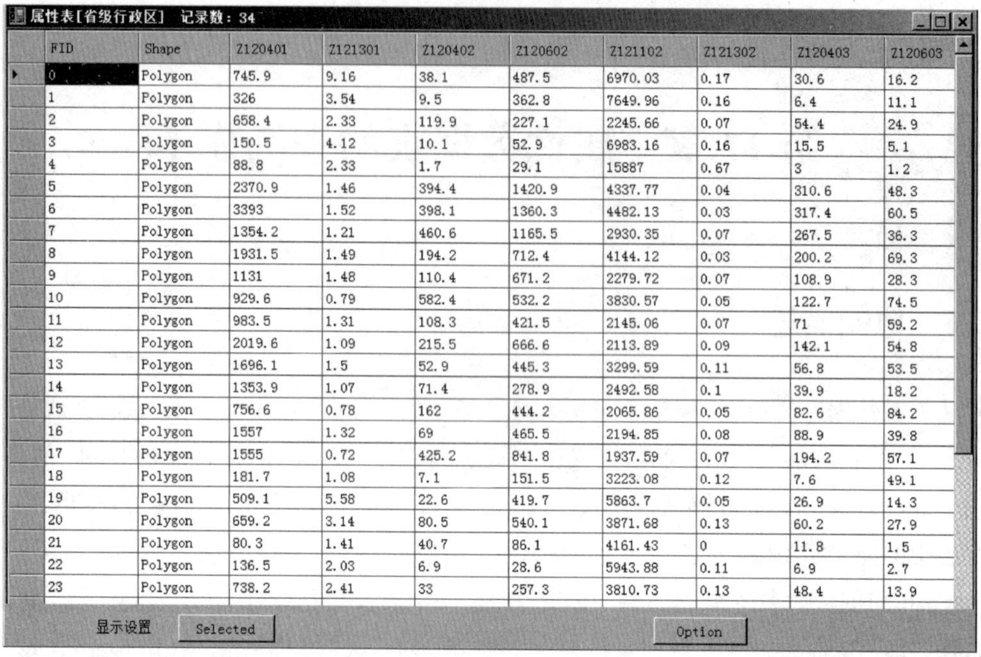

属性表[省级行政区]　记录数：34

FID	Shape	Z120401	Z121301	Z120402	Z120602	Z121102	Z121302	Z120403	Z120603
0	Polygon	745.9	9.16	38.1	487.5	6970.03	0.17	30.6	16.2
1	Polygon	326	3.54	9.5	362.8	7649.96	0.16	6.4	11.1
2	Polygon	658.4	2.33	119.9	227.1	2245.66	0.07	54.4	24.9
3	Polygon	150.5	4.12	10.1	52.9	6983.16	0.16	15.5	5.1
4	Polygon	88.8	2.33	1.7	29.1	15887	0.67	3	1.2
5	Polygon	2370.9	1.46	394.4	1420.9	4337.77	0.04	310.6	48.3
6	Polygon	3393	1.52	398.1	1360.3	4482.13	0.03	317.4	60.5
7	Polygon	1354.2	1.21	460.6	1165.5	2930.35	0.07	267.5	36.3
8	Polygon	1931.5	1.49	194.2	712.4	4144.12	0.03	200.2	69.3
9	Polygon	1131	1.48	110.4	671.2	2279.72	0.07	108.9	28.3
10	Polygon	929.6	0.79	582.4	532.2	3830.57	0.05	122.7	74.5
11	Polygon	983.5	1.31	108.3	421.5	2145.06	0.07	71	59.2
12	Polygon	2019.6	1.09	215.5	666.6	2113.89	0.09	142.1	54.8
13	Polygon	1696.1	1.5	52.9	445.3	3299.59	0.11	56.8	53.5
14	Polygon	1353.9	1.07	71.4	278.9	2492.58	0.1	39.9	18.2
15	Polygon	756.6	0.78	162	444.2	2065.86	0.05	82.6	84.2
16	Polygon	1557	1.32	69	465.5	2194.85	0.08	88.9	39.8
17	Polygon	1555	0.72	425.2	841.8	1937.59	0.07	194.2	57.1
18	Polygon	181.7	1.08	7.1	151.5	3223.08	0.12	7.6	49.1
19	Polygon	509.1	5.58	22.6	419.7	5863.7	0.05	26.9	14.3
20	Polygon	659.2	3.14	80.5	540.1	3871.68	0.13	60.2	27.9
21	Polygon	80.3	1.41	40.7	86.1	4161.43	0	11.8	1.5
22	Polygon	136.5	2.03	6.9	28.6	5943.88	0.11	6.9	2.7
23	Polygon	738.2	2.41	33	257.3	3810.73	0.13	48.4	13.9

显示设置　Selected　Option

图 17-1　属性表

从工具箱拖一个 DataGridView 控件到窗体(变量名为 dataGridView1)，并将其 Dock 属性设置为“Fill”。类设计代码如下：

```
public partial classOpenAttributeTableFrm:Form
{
    ILayer m_pLayer=null;
    //构造函数
    publicOpenAttributeTableFrm(IFeatureLayer pFeatureLayer)
    {
        InitializeComponent();
        m_pLayer=pFeatureLayer;
    }

    //装载事件响应函数
    private void OpenAttributeTableForm_Load(object sender,Even-
tArgs e)
    {
        DataSourceBinding(ILayer player);
    }
```

```
    //若干功能函数
    public voidDataSourceBinding(ILayer player)
    public DataTable CreateFeatureAttrTable(ILayer pLayer,string tableName)

    //若干辅助函数
    private DataTable CreateEmptyDataTable(ITable pTable,string ta-
bleName)
    //……
}
```

2. 创建空 DataTable

CreateEmptyDataTable()函数负责创建空 DataTable。步骤是:

①首先从传入 ILayer 查询到 ITable，从 ITable 中的 Fileds 中获得每个 Field。

②再根据 Filed 创建 DataTable 的 DataColumn。

③若干行对象构成空 DataTable。实现函数如下:

```
/// 根据图层字段创建一个只含字段的空 DataTable
private DataTable CreateEmptyDataTable ( ITable pTable, string ta-
bleName)
{
    //初始化 DataTable 表
    DataTable pDataTable=new DataTable(tableName);

    IField pField=null;
    DataColumn pDataColumn;
    //根据每个字段的属性建立 DataColumn 对象
    for (int i=0; i < pTable.Fields.FieldCount; i++)
    {
        pField=pTable.Fields.get_Field(i);
        //新建一个 DataColumn
        bool bUnique=(pField.Name==pTable.OIDFieldName) ? true:false;
        pDataColumn=CreateDataColumnByField(pField,bUnique);

        //字段添加到表括中
        pDataTable.Columns.Add(pDataColumn);
        pField=null;
        pDataColumn=null;
    }
    return pDataTable;
}
```

这里 DataColumn 是由 CreateDataColumnByField(…)创建，源代码如下：

```
///根据 IField 创建一个 DataColumn
private DataColumn CreateDataColumnByField ( IField pField, bool
bUnique)
{
    DataColumn pDataColumn=new DataColumn(pField.Name);
    //字段值是否唯一
    pDataColumn.Unique=bUnique;

    //字段值是否允许为空
    pDataColumn.AllowDBNull=pField.IsNullable;
    //字段别名
    pDataColumn.Caption=pField.AliasName;
    //字段数据类型
    pDataColumn.DataType=System.Type.GetType(ParseFieldType(pField.
Type));
    //字段默认值
    pDataColumn.DefaultValue=pField.DefaultValue;
    //当字段为 String 类型时设置字段长度
    if (pField.VarType==8)
    {
        pDataColumn.MaxLength=pField.Length;
    }

    return pDataColumn;
}
```

因为 GeoDatabase 的数据类型与.NET 的数据类型不同，故要进行转换。转换函数如下：

```
///将 GeoDatabase 字段类型转换成.Net 相应的数据类型
public string ParseFieldType(esriFieldType fieldType)
{
    switch (fieldType)
    {
    case esriFieldType.esriFieldTypeBlob:
        return "System.String";
    case esriFieldType.esriFieldTypeDate:
        return "System.DateTime";
    case esriFieldType.esriFieldTypeDouble:
```

```
            return "System.Double";
        case esriFieldType.esriFieldTypeGeometry:
            return "System.String";
        case esriFieldType.esriFieldTypeGlobalID:
            return "System.String";
        case esriFieldType.esriFieldTypeGUID:
            return "System.String";
        case esriFieldType.esriFieldTypeInteger:
            return "System.Int32";
        case esriFieldType.esriFieldTypeOID:
            return "System.String";
        case esriFieldType.esriFieldTypeRaster:
            return "System.String";
        case esriFieldType.esriFieldTypeSingle:
            return "System.Single";
        case esriFieldType.esriFieldTypeSmallInteger:
            return "System.Int32";
        case esriFieldType.esriFieldTypeString:
            return "System.String";
        default:
            return "System.String";
    }
}
```

3. 将要素类转换为 DataTable 数据

TransformFeatureAttrTable( )函数将要素类转换为 DataTable 数据，实现步骤是：

①创建空 DataTable；

②通过 ICursor 遍历 ITable 行(即 IRow)，每一行创建 DataTable 中相应的 DataRow；

③再将所有的 DataRow 添加到 DataTable 中。

为保证效率，一次最多只装载 2000 条数据到 DataGridView。函数代码如下：

```
///生成 DataTable 中的数据
public DataTable TransformFeatureAttrTable(IFeatureLayer pLayer,
string tableName)
{
    //创建空 DataTable
    ITable pTable=pLayer as ITable;
    DataTable pDataTable=CreateEmptyDataTable(pTable,tableName);
    //取得图层类型
    string shapeType=getShapeType(pLayer);
```

```
//创建 DataTable 的行对象
DataRow pDataRow=null;

//创建查询游标;
ITable pTable=pLayer as ITable;
ICursor pCursor=pTable.Search(null,false);

//取得 ITable 中的行信息
IRow pRow=pCursor.NextRow();
int n=0;
while (pRow!=null)
{
    //新建 DataTable 的行对象
    pDataRow=pDataTable.NewRow();
    for (int i=0; i < pRow.Fields.FieldCount; i++)
    {
        //如果字段类型为 esriFieldTypeGeometry,则根据图层类型设置字段值
        if (pRow.Fields.get_Field(i).Type ==
                        esriFieldType.esriFieldTypeGeometry)
        {
            pDataRow[i]=shapeType;
        }
        //当图层类型为 Anotation 时,要素类中会有 esriFieldTypeBlob 类型数据,
        //其存储的是标注内容,如此情况需将对应的字段值设置为 Element
        else if (pRow.Fields.get_Field(i).Type ==
                            esriFieldType.esriFieldTypeBlob)
        {
            pDataRow[i]="Element";
        }
        else
        {
            pDataRow[i]=pRow.get_Value(i);
        }
    }
    //添加 DataRow 到 DataTable
    pDataTable.Rows.Add(pDataRow);
    pDataRow=null;
    n++;
```

```
            //为保证效率,一次只装载最多条记录
            if (n==2000)
            {
                pRow=null;
            }
            else
            {
                pRow=pCursor.NextRow();
            }
        }
        return pDataTable;
    }
```

上面的代码中将 esriFieldTypeGeometry 字段用几何类型名表示，涉及一个获取图层几何类型的函数 getShapeTape，代码如下：

```
    ///获得图层的 Shape 类型
    public string getShapeType(ILayer pLayer)
    {
        IFeatureLayer pFeatLyr=(IFeatureLayer)pLayer;
        switch (pFeatLyr.FeatureClass.ShapeType)
        {
        case esriGeometryType.esriGeometryPoint:
            return "Point";
        case esriGeometryType.esriGeometryPolyline:
            return "Polyline";
        case esriGeometryType.esriGeometryPolygon:
            return "Polygon";
        default:
            return "";
        }
    }
```

4. 绑定 DataTable 到 DataGridView

通过以上步骤，我们已经得到了一个含有图层属性数据的 DataTable。通过 DataSourceBinding()函数，我们很容易将其绑定到 DataGridView 控件中。

```
    /// 绑定 DataTable 到 DataGridView
    public voidDataSourceBinding(ILayer player)
    {
        string tableName=getValidFeatureClassName(player.Name);
        DataTable attributeTable=TransformFeatureAttrTable(player,ta-
```

```
bleName );

        this.dataGridView1.DataSource=attributeTable;
        this.Text="属性表["+tableName+"] "+"记录数:"+
                attributeTable.Rows.Count .ToString();
    }
```

因为 DataTable 的表名不允许含有“.”，因此我们用“_”替换。函数如下：

```
///替换数据表名中的“.”符号
public string getValidFeatureClassName(string FCname)
{
    int dot=FCname.IndexOf(".");
    if (dot!=-1)
    {
        return FCname.Replace(".","_");
    }
    return FCname;
}
```

17.4 功能调用

通过以上步骤，我们封装了一个 OpenAttributeTableFrm 类，此类能够由 FeatureLayer 显示图层中的属性表数据。

在图层操作浮动菜单上添加菜单项——OpenAttributeTable。建立 OpenAttributeTable 的 OnClick 事件的响应函数，代码如下：

```
    private void openAttributeTableToolStripMenuItem_Click(object sender,
EventArgs e)
    {
        OpenAttributeTableForm frm = new OpenAttributeTableForm ( m _
tocRightLayer);
        frm.ShowDialog();
    }
```

17.5 编译运行

按下 F5 键，编译运行程序，此时已经实现了开篇处展示的属性表效果。

以上代码适用于 Windows 10+VS2015+AE10. 5 编译环境。

17.6　功能增强

17.6.1　选择集和全要素显示切换

以上源代码只支持全部要素属性显示，下面通过一个按钮 btnVicwMethod 实现选择集显示和全要素显示两种情形的切换，控制代码如下：

①在 Form 中底部添加一个 Panel(Dock 属性设为 Botton)，然后在 Panel 上添加 Button 按钮——btnViewMethod;

②添加类成员变量 m_isSelectionSet，该变量为真，显示选择集，否则显示全部要素。

③查询游标由函数 CreateCursorBySelectionSet 创建，此函数根据布尔变量 m_isSelectionSet 的值决定使用选择集创建游标，还是全要素创建游标，代码如下：

```
/// 从 Layer 查询到 Cursor
private ICursor CreateCursorBySelectionSet(ILayer pLayer)
{
    ICursor pCursor=null;
    if (m_isSelectionSet)
    {
        IFeatureSelection pSeletion=pLayer as IFeatureSelection;
        ISelectionSet pSelectionSet=pSeletion.SelectionSet;
        pSelectionSet.Search(null,false,out pCursor);
    }
    else
    {
        ITable pTable=pLayer as ITable;
        pCursor=pTable.Search(null,false);
    }
    return pCursor;
}
```

④将 TransformFeatureAttrTable()函数中创建的查询游标的两行代码;

```
ITable pTable=pLayer as ITable;
ICursor pCursor=pTable. Search(null, false);
```

改为:

```
ICursor pCursor=CreateCursorBySelectionSet( pLayer );
```

⑤添加 btnViewMethod 响应函数，代码如下(使得按压该按钮，DataGridView 显示内容在选择集和全要素来回切换)：

```
private void btnViewMethod_Click(object sender,EventArgs e)
{
```

```
        if (m_isSelectionSet)
        {
            DataSourceBinding(m_pLayer);
            m_isSelectionSet=false;
            btnViewMethod.Text="All";
        }
        else
        {
            DataSourceBinding(m_pLayer);
            m_isSelectionSet=true;
            btnViewMethod.Text="Selected";
        }
    }
```

17.6.2 栅格数据属性显示

栅格图层属性表显示与矢量数据显示的主要区别在于 DataGridView 的数据源不同，即只需要将 DataSource 捆绑为由栅格数据属性建立的 Datatable 即可。

①先建立 TransformRatserAttrTable 函数，源代码如下：

```
    public DataTable TransformRatserAttrTable(IRasterLayer pRlyr,string
tableName)
    {
        DataTable pTable=new DataTable();

        IRaster pRaster=pRlyr.Raster;
        IRasterProps pProp=pRaster as IRasterProps;
        if (pProp.PixelType==rstPixelType.PT_LONG)
        {
            IRasterBandCollection pBcol=pRaster as IRasterBandCollection;
            IRasterBand pBand=pBcol.Item(0);

            //判断是否存在属性表
            bool bAttrTableExist;
            pBand.HasTable(out bAttrTableExist);
            if (!bAttrTableExist)
                return null;

            //创建空表
            ITable pRTable=pBand.AttributeTable;
```

```
            pTable=CreateEmptyDataTable(pRTable,tableName);

            ICursor pCursor=pRTable.Search(null,false);
            IRow pRrow=pCursor.NextRow();
            while (pRrow!=null)
            {
                DataRow pRow=pTable.NewRow();
                for (int i=0; i < pRrow.Fields.FieldCount; i++)
                {
                    pRow[i]=pRrow.get_Value(i).ToString();
                }
                pTable.Rows.Add(pRow);
                pRrow=pCursor.NextRow();
            }
        }
        return pTable;
    }
```

②修改 DataSourceBinding 函数，判断图层的类型，调用对应的数据源创建函数，源代码如下：

```
    ///绑定 DataTable 到 DataGridView
    public void DataSourceBinding(ILayer player)
    {
        string tableName=getValidFeatureClassName(player.Name);
        DataTable attributeTable=null;
        if (player is IFeatureLayer)
        {
            attributeTable=TransformFeatureAttrTable(player as IFeature-
Layer,
                                                                  tableName);
            this.AttrGridView.DataSource=attributeTable;
            this.Text="属性表["+tableName+"]"+"记录数:"+
                        attributeTable.Rows.Count.ToString();
        }
        else if (player is IRasterLayer)
        {
            attributeTable=TransformRatserAttrTable(player as IRaster-
Layer,tableName);
```

```
            this.AttrGridView.DataSource=attributeTable;
            string strCount=(attributeTable!=null) ?
                                attributeTable.Rows.Count.ToString():"属
性表不存在";
            this.Text="属性表["+tableName+"] "+"记录数:"+strCount;
                          ;
        }
    }
```

③修改 Form 加载响应函数，如果图层为栅格数据，使 btnViewMethod 无效，代码如下：

```
    private void OpenAttributeTableForm_Load(object sender,EventArgs e)
    {
        if (!(m_pLayer is IFeatureLayer))
        {
            this.btnViewMethod.Invalidate();
        }
        else
        {
            isSelectionSet=false;
            btnViewMethod.Text="All";
        }
        btnViewMethod_Click(sender,e);
    }
```

17.6.3 添加浮动式功能菜单

1. 添加浮动菜单

为进一步增强功能，共添加三个浮动菜单：

①单击 Option 按钮，弹出一个浮动菜单，包括 AddField(添加字段)、Export(导出)、Print(打印)等菜单项。

②右键单击任意列头，弹出一个浮动菜单，包括 DeleteField(删除字段)，Calculate(字段计算)，Sorting(字段排序)，Statistics(统计计算)等菜单项；可通过响应 DataGridView 的 ColumnHeaderMouseClick 事件弹出。

③右键单击任意行头，弹出一个浮动菜单，包括 Remove(删除行)，Selected(选择该行)等菜单项；可通过响应 DataGridView 的 RowHeaderMouseClick 事件弹出。

ColumnHeaderMouseClick 和 RowHeaderMouseClick 的响应函数代码如下：

```
    private void AttrGridView_ColumnHeaderMouseClick(object sender,
DataGridViewCellMouseEventArgs e)
    {
```

```
        if (e.Button==System.Windows.Forms.MouseButtons.Right)
        {
            m_columnSelectedIndex=e.ColumnIndex;
            System.Drawing.Point _point=
            this.dataGridView1.PointToClient(System.Windows.Forms.Cursor.Position);

            this.contextMenuStripColumn.Show(this.AttrGridView,_point);
        }
    }
    private void AttrGridView_RowHeaderMouseClick(object sender,DataGridViewCellMouseEventArgs e)
    {
        if (e.Button==System.Windows.Forms.MouseButtons.Right)
        {
            m_rowSelectedIndex=e.RowIndex;
            this.contextMenuStripRow.Show(MousePosition.X,MousePosition.Y);
        }
    }
```

注意：为记录鼠标点击的列号/行号，定义两个私有成员：

private intm_columnSelectedIndex=-1;

private intm_rowSelectedIndex=-1;

2. 菜单项响应函数

(1)Selected 菜单项

右键单击某行头，再单击 Selected 菜单项，该行所代表的要素被选中，方法是：根据对象 ID 所在的列和选定行索引，获取选定行的对象 ID，然后以此构造查询条件用 IFeatureSelection 接口进行选择操作，最后发送一个消息通知 MapControl 进行刷新，选中要素高亮显示，代码如下：

```
    private void selectedToolStripMenuItem_Click(object sender,EventArgs e)
    {
        ITable pTable=m_pLayer as ITable;
        IFields pFields=pTable.Fields;
        //获取对象 ID 所在列的索引
        int oidIndex=pFields.FindField(pTable.OIDFieldName);
        //获取选定行对象 ID 的值;
        IRow pRow=pTable.GetRow(m_rowSelectedIndex);
        string idString=pRow.get_Value(oidIndex).ToString();
```

```
        //建立查询过滤器
        IQueryFilter pFilter=new QueryFilterClass();
        pFilter.WhereClause=pTable.OIDFieldName +" ="+idString;
        //执行选择操作;
        IFeatureSelection pSel=m_pLayer as IFeatureSelection;
        pSel.SelectFeatures(pFilter, esriSelectionResultEnum.esriSe-
lectionResultNew,true);
        //通知 MapControl 刷新
        ForceMapControlRefreshEvent ("","");
    }
```

这里 ForceMapControlRefreshEvent ("","")通知 MapControl 刷新，需要用到自定义委托 delegate 和事件 Event，做法如下：

①在 OpenAttributeTableForm 添加 Event 和委托 delegate，代码如下：

```
    public delegate void NotifyMapControlRefreshEventHandler ( string
sFeatClsName,
                                                               string
sFieldName);
    public event NotifyMapControlRefreshEventHandler ForceMapControl-
RefreshEvent=null;
```

②调用时订阅 NotifyMapControlRefreshEventHandler 事件，代码如下：

```
    private void openAttributeTableToolStripMenuItem _ Click ( object
sender,EventArgs e)
    {
        OpenAttributeTableForm frm=new OpenAttributeTableForm(m_tocRightLa-
yer);
        frm .ForceMapControlRefreshEvent +=new
                OpenAttributeTableForm. NotifyMapControlRefreshEventHandler
(MapControl_Refresh);

        frm .ShowDialog();
    }
```

③响应函数刷新地图，代码如下：

```
    private void MapControl_Refresh(string sFeatClsName,string sField-
Name)
    {
        this._mapControl.ActiveView.PartialRefresh(esriViewDrawPhase.es-
riViewAll,null,null);
    }
```

(2) DeleteField 菜单项

右键点击某列头，再点击 DeleteField 菜单项，删除点中的列：用 ITable 的 DeleteField 方法删除数据集的对应字段，同时在 DataGridView 中移除选中的列，代码如下：

```
    private void deleteFieldToolStripMenuItem_Click(object sender,
EventArgs e)
    {
        ITable pTable=m_pLayer as ITable;
        IFields pFields=pTable.Fields;
        //查找要删除的字段
        IField pField=pFields.get_Field(m_columnSelectedIndex);
        //删除字段
        pTable.DeleteField(pField);

        //在 DataGridView 数据源移除删去的列
        DataTable pDtTable = (this.AttrGridView.DataSource) as DataT-
able;
        DataColumn pDataColumn = pDtTable.Columns[m_columnSelectedIn-
dex];
        pDtTable.Columns.Remove(pDataColumn);
    }
```

(3) AddField 菜单项

点击 Option 按钮，再点击 AddField 菜单项，添加新字段：先创建一个新字段 pField，用 ITable 的 AddField 方法添加新的字段到数据集，同时在 DataGridView 中添加新列，代码如下：

```
    private void addFieldToolStripMenuItem_Click(object sender,Even-
tArgs e)
    {
        //建立字段 Greencover
        IField pField=new FieldClass();
        IFieldEdit pFieldEdit;
        pFieldEdit=pField as IFieldEdit;
        pFieldEdit.Type_2=esriFieldType.esriFieldTypeDouble;
        pFieldEdit.Name_2="Greencover";

        ITable pTable=m_pLayer as ITable;
        if (pTable.Fields.FindField(pField.Name) < 0 )
        {
            pTable.AddField(pField);
```

```
            //在 DataGridView 数据源加新增列
            DataTable pDtTable=(this.AttrGridView.DataSource) as DataTable;
            DataColumn pDataColumn=CreateDataColumnByField(pField,false);
            pDtTable.Columns.Add(pDataColumn);
        }
    }
```

(4)Caculate 菜单项

右键点击某列头，再点击 Caculate 菜单项，计算该列所有值(这里以面积计算为例)，同时更新要素类和在 DataGridView 中的列信息，代码如下：

```
    private void caculateToolStripMenuItem_Click(object sender,Even-
tArgs e)
    {
        IFeatureClass pFcls =(m_pFeatureLayer as IFeatureLayer).Fea-
tureClass;
        IFeatureCursor pfCursor=pFcls.Search(null,false);
        IFeature pFt=null;
        while ((pFt=pfCursor.NextFeature())!=null)
        {
            //通过 IArea 更新面积字段;
            IArea pArea=pFt.Shape as IArea;
            double area=pArea.Area;
            pFt.set_Value(m_columnSelectedIndex,area);
            pFt.Store(); //存储变化
        }
    }
```

(5)Statistics 菜单项

右键点击某列头，再点击 Statistics 菜单项，计算该列统计信息，代码如下：

```
    private void statisticsToolStripMenuItem_Click(object sender,Even-
tArgs e)
    {
        //创建游标(结果只有一个字段)
        IFeatureClass pFcls=(m_pLayer as IFeatureLayer).FeatureClass;
        IFeatureCursor pCursor=pFcls.Search(null,true);
        IField pField=pFcls.Fields.Field[m_columnSelectedIndex];
        //创建数据统计对象
        IDataStatistics pDastStat=new DataStatistics();
        {
            pDastStat.Field=pField.Name;
```

```
            pDastStat.Cursor=(ICursor)pCursor;
        }
        IStatisticsResults pStatistics=pDastStat.Statistics;
        string strMean="Mean="+pStatistics.Mean.ToString()+Environ-
ment.NewLine;
        string strMax="Max="+pStatistics.Maximum.ToString()+Environ-
ment.NewLine;
        string strMin="Min="+pStatistics.Minimum.ToString()+Environ-
ment.NewLine;
        string strSum="Sum="+pStatistics.Sum.ToString();
        MessageBox.Show(strMean+strMax+strMin+strSum);
    }
```

第18章 拓扑检查

18.1 知识要点

拓扑是一个或多个 Geodatabase 要素类的集合，这个集合共同遵守若干几何特征一致性规则(拓扑规则)。例如，公共道路、成片土地和建筑物遵守如下规则：地块不应重叠，地块和公共道路也不应重叠、建筑物应当被包含在地块内。

拓扑结构是隶属于数据集，一个要素数据集可以拥有多个拓扑，一个拓扑可以添加多个要素类，参与构建拓扑的要素应是简单要素类，拓扑不会修改要素类的定义。

ArcEngine 允许定义这些类型的规则，并提供验证工具来识别违反规则的功能，以及明确允许违反规则的功能。常用接口是 ITopologyContainer2/ ITopologyContainer，ITopology，ITopologyRuleContainer 接口：

①ITopologyContainer2 用于创建 ITopology，用到如下两个方法：ITopologyContainer. CreateTopology( )和 ITopologyContainer2. CreateTopologyEx( )。

②ITopologyRuleContainer 添加拓扑规则；TopologyRuleContainer. AddRule( )。

③ITopology 添加图层，验证拓扑：ITopology. ValidateTopology( )和 ITopology. AddClass( )。

18.2 功能描述

点击【Data Manager】Tab 页的【Topology】按钮，弹出拓扑分析对话框：即根据选定的目标数据集，在列表中定义拓扑规则等，即可进行拓扑检查分析。操作界面如图 18-1 所示。

图 18-1 拓扑检查分析对话框

18.3　功能实现

18.3.1　新建功能窗体

1. 界面设计

项目中添加一个新的窗体，名称为“TopologyAanylysisFrm”，Name 属性设为“Topology Aanylysis”，添加 1 个 ComboBox、1 个 TextBox、4 个 Button 控件，1 个 DataGridView。

控件属性设置见表 18-1。

表 18-1　　控件属性

控件类型	Name 属性	控件说明	备　注
ComBox	cbxDatasetName	目标数据集	
TextBox	txtTopoName	拓扑检查名称	
Button	btnDatasetBrowesr	数据集浏览	
Button	btnTopoAnalyst	拓扑分析	
Button	btnTopoValidated	拓扑验证	
Button	btnCancel	取消	
DataGridView	dataGridView1	拓扑定义信息表	

2. 类结构设计

添加如下引用代码，修改类定义代码：

```
public partial class TopologyAanylysisFrm:Form
{
    private IWorkspace m_pWorkspace=null;
    public TopologyAanylysisFrm()
    {
        InitializeComponent();
    }

    #region 消息响应函数
    private void TopologyAanylysisFrm_Load(object sender,EventArgs e)
    private void btnDatasetBrowesr_Click(object sender,EventArgs e)
    private void cbxDatasetName_SelectedIndexChanged(object sender,Even-
tArgs e)
    private void btnTopologyCheck_Click(object sender,EventArgs e)
```

```
        private void btnTopologyValidated_Click(object sender,EventArgs e)
        #endregion

        #region 核心功能函数
        //拓扑检查
        public void TopologyCheck()
        //创建新拓扑
        public ITopology CreateTopology(IFeatureDataset featureDataset,
string sTopologyName,double dTolerance,bool specifyZClusterTolerance)
        //添加拓扑规则
        public void AddRuleToTopology(ITopology topology,esriTopology-
RuleType ruleType,String ruleName,IFeatureClass featureClass)
        public void AddRuleToTopology(ITopology topology,esriTopology-
RuleType ruleType, String ruleName, IFeatureClass originClass, int ori-
ginSubtype,IFeatureClass destinationClass,int destinationSubtype)
        //拓扑验证
        public void ValidateTopology(ITopology topology,IEnvelope enve-
lope)
        #endregion

        #region 辅助函数
        private esriTopologyRuleType getTopologyRuleType(string ruleName)
        private bool IsDoubleDataTopoRule(string ruleName)
        private DataGridViewComboBoxColumn CreateComboBoxWithEnums(string
Title)
        #endregion
    }
```

18.3.2 消息响应函数

1. 载入响应函数 TopologyAanylysisFrm_Load()

TopologyAanylysisFrm 在载入时，主要完成 dataGridView1 初始化，共构建 3 个 ComboBox 类型的列元素，分别是“拓扑规则”，“参与数据”，“参与数据+”。其中，“拓扑规则”列元素填充 5 个拓扑规则名：esriTRTLineNoDangles(无悬挂点)，esriTRTLineNoPseudos(无伪节点)，esriTRTAreaNoOverlap(面无重叠)，esriTRTAreaCoveredByAreaClass(面无覆盖)，esriTRTPointCoveredByAreaBoundary(点不能被面覆盖)。

代码如下：

```
    private void TopologyAanylysisFrm_Load(object sender,EventArgs e)
    {
```

```
        DataGridViewComboBoxColumn dgvCbxRule = CreateComboBoxWithEnums("
拓扑规则");
        dgvCbxRule.Items.Add(esriTopologyRuleType.esriTRTLineNoDangles.
ToString());
        dgvCbxRule.Items.Add(esriTopologyRuleType.esriTRTLineNoPseudos.
ToString());
        dgvCbxRule.Items.Add(esriTopologyRuleType.esriTRTAreaNoOver-
lap.ToString());
        dgvCbxRule.Items.Add(esriTopologyRuleType.esriTRTAreaCovered-
ByAreaClass.ToString());
        dgvCbxRule.Items.Add(esriTopologyRuleType.esriTRTPointCovered-
ByAreaBoundary.ToString());

        DataGridViewComboBoxColumn dgvCbxInFClass = CreateComboBoxWith-
Enums("参与数据");
        dgvCbxInFClass.Items.Add("NULL");
        DataGridViewComboBoxColumn dgvCbxrefFClass = CreateComboBoxWithE-
nums("参与数据+");
        dgvCbxrefFClass.Items.Add("NULL");

        this.dataGridView1.Columns.Add(dgvCbxInFClass);
        this.dataGridView1.Columns.Add(dgvCbxRule);
        this.dataGridView1.Columns.Add(dgvCbxrefFClass);
    }
```

2. 数据集设置响应函数 btnDatasetBrowser_Click()

本例中数据集设置 gdb 数据源，步骤如下：

①通过 FolderBrowserDialog 对话框选择 gdb 数据源目录；

②打开 gdb 工作空间；

③然后通过 IEnumDatasetName 名字接口将 gdb 中所有的数据集添加到 cbxDatasetName 的 Items 中，代码如下：

```
    private void btnDatasetBrowesr_Click(object sender,EventArgs e)
    {
        FolderBrowserDialog folderDlg = new FolderBrowserDialog();
        DialogResult dr = folderDlg.ShowDialog();
        if (dr = = DialogResult.OK)
        {
            //选择 gdb 数据源目录
            string txtDatabaseName = folderDlg.SelectedPath;
```

```
            Type factoryType =
                          Type.GetTypeFromProgID("esriDataSourcesGDB.Fi-
leGDBWorkspaceFactory");

            //打开工作空间
            IWorkspaceFactory workspaceFactory = null;
            workspaceFactory = (IWorkspaceFactory)Activator.CreateIn-
stance(factoryType);
            m_pWorkspace =workspaceFactory.OpenFromFile(txtDatabaseName,0);

            //获取 Dataset 名称枚举接口变量
            IEnumDatasetNamedatasetName = null;
            datasetName = m_pWorkspace.get_DatasetNames(esriDatasetType.
esriDTFeatureDataset);

            //填充 cbxDatasetName
            IDatasetName dtName =null;
            while ((dtName = datasetName.Next())! =null)
            {
                this.cbxDatasetName.Items.Add(dtName.Name);
            }

                this.cbxDatasetName.SelectedIndex = 0;
        }
    }
```

3. 数据集名称索引变化响应函数 cbxDatasetName_SelectedIndexChanged( )

①根据数据集名查找数据集名字接口 IDatasetName;

②清空 DataGridView 的 0/2 列 ComboBox;

③用数据集名称接口的子集名填充 DataGridView 的 0/2 列 ComboBox。

代码如下:

```
    private void cbxDatasetName_SelectedIndexChanged(object sender,
EventArgs e)
    {
        //根据数据集名查找数据集名称接口:
        IEnumDatasetName eDatasetName = null;
        eDatasetName = m_pWorkspace.get_DatasetNames(esriDatasetType.
esriDTFeatureDataset);
        IDatasetName dtName = null;
```

```
        while ((dtName=eDatasetName.Next())!=null)
        {
            if(dtName.Name==cbxDatasetName.SelectedItem.ToString())
                break;
        }

        //清空 DataGridView 的 ComboBox 列(0,2);
        DataGridViewComboBoxColumn dgvCbxrefFClass=null;
        DataGridViewComboBoxColumn dgvCbxInFClas=null;
        dgvCbxrefFClass =  this.dataGridView1.Columns[0] as DataGrid-
ViewComboBoxColumn;
        dgvCbxInFClass=this.dataGridView1.Columns[2] as DataGridView-
ComboBoxColumn;
        dgvCbxrefFClass.Items.Clear();
        dgvCbxInFClass.Items.Clear();

        //用数据集名称接口的子集填充 DataGridView 的 ComboBox 列(0,2);
        IEnumDatasetName eSubsetName=dtName.SubsetNames;
        IDatasetName subDatName=null;
        while ((subDatName=eSubsetName.Next())!=null)
        {
            dgvCbxrefFClass.Items.Add(subDatName.Name);
            dgvCbxInFClass.Items.Add(subDatName.Name);
        }
    }
```

4. 拓扑检查/拓扑验证响应函数

此处两个响应函数：btnTopologyCheck_ Click()/ btnTopologyValidated_ Click()分别调用 TopologyCheck()，ValidateTopology() 核心功能函数：

```
    private void btnTopologyCheck_Click(object sender,EventArgs e)
    {
        TopologyCheck();
    }

    private void btnTopologyValidated_Click(object sender,EventArgs e)
    {
        //打开拓扑
        string featureDatasetName=this.cbxDatasetName.Text;
        string topologyName=this.txtTopoName.Text;
```

```
        ITopology topology = OpenTopologyByName(featureDatasetName,to-
pologyName);

        //无边际范围
        IEnvelope envelope = new EnvelopeClass();
        envelope.PutCoords(double.MinValue,double.MinValue,double.
MaxValue,double.MaxValue);

        //执行验证操作
        ValidateTopology(topology,envelope);
    }
```

18.3.3 核心函数

1. TopologyCheck( )

①使用 ISchemaLock 的 ChangeSchemaLock 函数，在要素数据集上建立独占模式锁；

②创建拓扑：此处使用核心函数 CreateTopology( )；

③添加要素类和拓扑规则到拓扑结构中：每次根据拓扑规则涉及的要素类多少，使用不同的 AddRuleToTopology 函数；

④调用 ValidateTopology( )进行验证拓扑；

⑤最后将数据集上独占模式锁改为共享模式。代码如下：

```
    public void TopologyCheck()
    {
        // Open the workspace and the required datasets.
        IFeatureWorkspace featureWorkspace = (IFeatureWorkspace) m_
pWorkspace;
        string strDatasetName = cbxDatasetName.SelectedItem.ToString();
        IFeatureDataset featureDataset = featureWorkspace.OpenFeature-
Dataset(strDatasetName);

        ISchemaLock schemaLock = (ISchemaLock)featureDataset;
        try
        {
          //尝试在要素数据集上建立独占模式锁
            schemaLock.ChangeSchemaLock(esriSchemaLock.esriExclusive-
SchemaLock);

            //创建拓扑
            string topoName = this.cbxDatasetName.SelectedItem.ToString();
```

```
            topoName += "_"+this.txtTopoName.Text.ToString();
            ITopology topology=CreateTopology(featureDataset,topoName,-1,
false);

            //添加要素类和拓扑规则到拓扑结构中
            for (int i=0; i < this.dataGridView1.RowCount - 1; i++)
            {
                string ruleName=dataGridView1[1,i].Value.ToString();
                esriTopologyRuleType ruleType=getTopologyRuleType(ruleName);

                if (!IsDoubleDataTopoRule(ruleName) )
                {
                     string inFClassName=dataGridView1[0,i].Value.ToS-
tring();
                      IFeatureClass blocksFC = featureWorkspace.OpenFea-
tureClass(inFClassName);

                    topology.AddClass(blocksFC,5,1,1,false);
                    AddRuleToTopology(topology,ruleType,ruleType.ToString
(),blocksFC);
                }
                else
                {
                    string inFClassName=dataGridView1[0,i].Value.ToString();
                    string refFClassName=dataGridView1[2,i].Value.ToS-
tring();
                    IFeatureClass blocksFC = featureWorkspace.OpenFeature-
Class(inFClassName);
                    IFeatureClass parcelsFC=featureWorkspace.OpenFeature-
Class(refFClassName);

                    topology.AddClass(blocksFC,5,1,1,false);
                    topology.AddClass(parcelsFC,5,1,1,false);
                    AddRuleToTopology(topology,ruleType,ruleType.ToString
(),parcelsFC,1,
                                        blocksFC,1);
                }
            }
```

```
            //获取验证拓扑的范围并且验证拓扑
            IGeoDatasetgDataset=(IGeoDataset)topology;
            IEnvelope envelope=gDataset.Extent;
            ValidateTopology(topology,envelope);
        }
        catch (COMException comExc)
        {
            MessageBox.Show(comExc.Message.ToString());
        }
        finally
        {
            schemaLock.ChangeSchemaLock(esriSchemaLock.esriSharedSchemaLock);
        }
    }
```

2. AddRuleToTopology()函数

单要素拓扑规则使用第一个 AddRuleToTopology 函数，双要素拓扑规则使用第二个 AddRuleToTopology 函数，代码如下：

```
    //添加拓扑规则第一函数
    public void AddRuleToTopology(ITopology topology,esriTopologyRule-
Type ruleType,String ruleName,IFeatureClass featureClass)
    {
        //初始化拓扑规则
        ITopologyRule topologyRule=new TopologyRuleClass();
        topologyRule.TopologyRuleType=ruleType;
        topologyRule.Name=ruleName;
        topologyRule.OriginClassID=featureClass.FeatureClassID;
        topologyRule.AllOriginSubtypes=true;

        //把 topology 对象强制转换到 ITopologyRuleContainer 对象,然后添加拓扑规则
        ITopologyRuleContainer topologyRuleContainer=(ITopologyRule-
Container)topology;
        if (topologyRuleContainer.get_CanAddRule(topologyRule))
        {
            topologyRuleContainer.AddRule(topologyRule);
        }
        else
```

```
        {
            throw new ArgumentException("Could not add specified rule to
the topology.");
        }
    }

    //添加拓扑规则第二函数
    public void AddRuleToTopology(ITopology topology,esriTopologyRule-
Type ruleType,String ruleName,IFeatureClass originClass,int originSub-
type,IFeatureClass destinationClass,int destinationSubtype)
    {
        //初始化拓扑规则
        ITopologyRule topologyRule=new TopologyRuleClass();
        topologyRule.TopologyRuleType=ruleType;
        topologyRule.Name=ruleName;
        topologyRule.OriginClassID=originClass.FeatureClassID;
        topologyRule.AllOriginSubtypes=false;
        topologyRule.OriginSubtype=originSubtype;
        topologyRule.DestinationClassID=destinationClass.FeatureClassID;
        topologyRule.AllDestinationSubtypes=false;
        topologyRule.DestinationSubtype=destinationSubtype;

        // Cast the topology to the ITopologyRuleContainer interface and
add the rule.
        ITopologyRuleContainer topologyRuleContainer = (ITopologyRule-
Container)topology;
        if (topologyRuleContainer.get_CanAddRule(topologyRule))
        {
            topologyRuleContainer.AddRule(topologyRule);
        }
        else
        {
            throw new ArgumentException("Could not add specified rule to
the topology.");
        }
    }
```

3. ValidateTopology()函数

代码如下：

```
public void ValidateTopology(ITopology topology, IEnvelope envelope)
{
    // Get the dirty area within the provided envelope.
    IPolygon locationPolygon=new PolygonClass();
    ISegmentCollection segmentCollection=(ISegmentCollection)locationPolygon;
    segmentCollection.SetRectangle(envelope);
    IPolygon polygon=topology.get_DirtyArea(locationPolygon);

    // If a dirty area exists,validate the topology.
    if (!polygon.IsEmpty)
    {
        // Define the area to validate and validate the topology.
        IEnvelope areaToValidate=polygon.Envelope;
        IEnvelope areaValidated=topology.ValidateTopology(areaToValidate);
    }
}
```

4. OpenTopologyByName()函数

代码如下:

```
//打开拓扑
public ITopology OpenTopologyByName(String featureDatasetName, String topologyName)
{
    //打开数据集
    IFeatureWorkspace pFws=m_pWorkspace as IFeatureWorkspace;
    IFeatureDataset featureDataset=pFws.OpenFeatureDataset(featureDatasetName);
    //获取拓扑容器
    ITopologyContainer topologyContainer=(ITopologyContainer)featureDataset;
    //打开拓扑
    ITopology topology=topologyContainer.get_TopologyByName(topologyName);
    return topology;
}
```

18.3.4　辅助函数

```
private esriTopologyRuleType getTopologyRuleType(string ruleName)
{
    esriTopologyRuleType type=esriTopologyRuleType.esriTRTAny;
    switch ( ruleName )
    {
        case "esriTRTLineNoDangles":
            type=esriTopologyRuleType.esriTRTLineNoDangles;
            break;
        case "esriTRTLineNoPseudos":
            type=esriTopologyRuleType.esriTRTLineNoPseudos;
            break;
        case "esriTRTAreaNoOverlap":
            type=esriTopologyRuleType.esriTRTAreaNoOverlap;
            break;
        case "esriTRTAreaCoveredByAreaClass":
            type=esriTopologyRuleType.esriTRTAreaCoveredByAreaClass;
            break;
        case "esriTRTPointCoveredByAreaBoundary":
            type = esriTopologyRuleType.esriTRTPointCoveredByAreaBoun-
dary;
            break;
    }

    return type;
}

private bool IsDoubleDataTopoRule(string ruleName)
{
    switch (ruleName)
    {
        case "esriTRTLineNoDangles":
        case "esriTRTLineNoPseudos":
        case "esriTRTAreaNoOverlap":
            return false;
            break;
        case "esriTRTAreaCoveredByAreaClass":
```

```
            case "esriTRTPointCoveredByAreaBoundary":
                return true;
                break;
            default:
                return false;
                break;
        }
    }

    private DataGridViewComboBoxColumn CreateComboBoxWithEnums(string
Title)
    {
        DataGridViewComboBoxColumn combo = new DataGridViewComboBoxCol-
umn();
        combo.FlatStyle = FlatStyle.Popup;
        combo.Name = Title;
        return combo;
    }
```

18.4 功能调用

在【Data Manager】Tab 页上，添加【Topology】按钮。建立 Click 响应函数：

```
private void btnTopology_Click(object sender,EventArgs e)
{
    TopologyAanylysisFrm frm = new TopologyAanylysisFrm();
    frm.ShowDialog();
}
```

18.5 编译测试

按下 F5 键，编译运行程序。

运行程序，点击按钮【Topology】，弹出分析窗口，添加拓扑规则，按压功能按钮即可完成拓扑检查与拓扑验证。

第 19 章　空间数据库访问

19.1　概述

Geodatabase 是 ArcGIS 存储空间地理信息的数据模型，可存储于关系型数据库管理系统或者文件系统(File gdb)，前者需要通过 ArcSDE 中间件技术作为操纵空间数据库的桥梁。Geodatabase 按数据集(IDataset)组织数据，或者说数据库中包含若干数据集，如要素数据集、栅格数据集，等等。要素类是要素数据集的子集(也可直接将要素类放到数据库，但不提倡)。

ArcEngine 通过工作空间(IWorkspace)访问 Geodatabase，工作空间可以看成是用户连接数据库的管道，它表达了包含一个或多个数据集的数据库或数据源(数据源可以是表、要素类、关系类等)，实现 IWorkspace 接口的类，通常还实现 IFeatureWorkspace，IRaster-Workspace，INetworkWorkspace 等接口，分别应对不同数据内容。

IWorkspace 必须由工作空间工厂(IWorkspaceFactory)创建，它是工作空间的发布者，允许客户通过一组连接属性打开特定的工作空间。连接属性用 IPropertySet 接口定义。ArcEngine 具有 IWorkspaceFactory 接口若干种实现类，以应对不同数据源。表 19-1 给出常用的数据源相应的工作空间工厂实现类。

表 19-1　**工作空间工厂实现类及其数据源**

序号	工作空间工厂实现类	数据源
1	AccessWorkspaccFactory	Microsoft Access 数据源
2	ExcclWorkspaceFactory	Excel 数据源
3	FileGDBWorkspaceFactory	Esri 专用文件数据库
4	InMemoryWorkspaceFactory	内存数据库
5	OLEDBWorkspaceFactory	OLEDB 数据库
6	SdeWorkspaceFactory	SDE 数据库
7	Cad WorkspaceFactory	CAD 数据源
8	ShapefileWorkspaceFactory	Shape 文件数据源

一旦建立工作空间，我们就可以用它打开数据源的数据集，创建新数据存储结构，等

等。利用数据集(要素类)接口可以对打开的数据集进行数据入库或更新操作。

本章以 ArcSDE 10.1 为例介绍 ArcSDE 安装、配置、连接和访问。安装环境如下：

测试数据库：Microsoft SQL Server 2016 Express；

操作系统：Windows 7 SP1(64 位)，机器名为 WIN-KG9LKA8CBST，注意关闭系统防火墙。

19.2 创建 SQLExpress 地理数据库

为简化学习曲线，这里使用不需要用户许可的 ArcSDE for SQL Server Express 进行介绍。操作步骤详见附录 1。希望创建 ArcSDE 企业级数据库的读者，需要 ArcGIS Server 用户许可，请参照本书附录 2 进行安装。

19.3 连接 GeoDatabase 数据库

使用 ArcEngine 连接 SDE 数据库，涉及 IWorkspaceFactory，IWorkspace，IPropertySet(SDE 空间连接属性)三个接口，我们先建立两个窗体：

①“SDE 连接”窗体，取名为：ConnectSdeServerFrm；

②数据库内容浏览窗体，取名为：DbManagerDockFrm，为方便后续操作，该窗体设计为停靠窗体(继承 DockContent)。

19.3.1 ConnectSdeServerFrm 实现

1. ConnectSdeServerFrm 设计

ConnectSdeServerFrm 界面设计如图 19-1 所示。

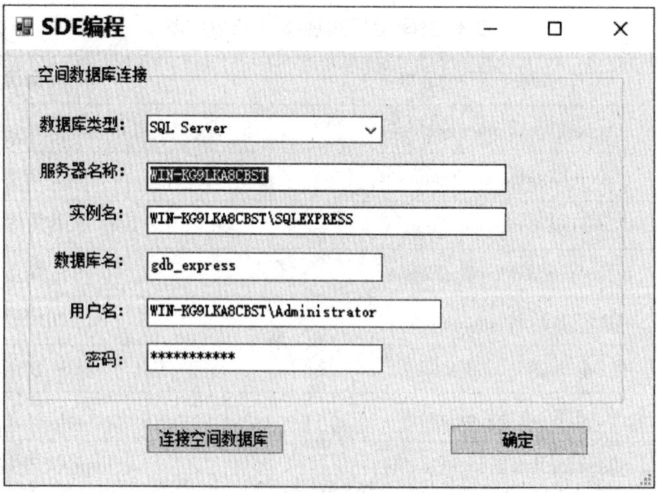

图 19-1 界面设计

2. 代码实现

具体代码如下：

```
/// <summary>
/// SDE 连接
/// </summary>
/// <param name="sKey"></param>
/// <returns></returns>
public partial classConnectSdeServerFrm:Form
{
    //工作空间
    public IWorkspace workspace;
    publicConnectSdeServerFrm()
    {
        workspace=null;
        InitializeComponent();
    }

    private void ConnectSdeServerFrm_Load(object sender,EventArgs e)
    {
        //IAoInitialize pao=new AoInitializeClass();
        //pao.Initialize(esriLicenseProductCode.esriLicenseProductCode
EngineGeoDB);
        this.cbxDatabaseType.SelectedIndex=0;
    }

    private void btnConnect_Click(object sender,EventArgs e)
    {
        string dbString=this.cbxDatabaseType.SelectedItem.ToString();
        string strFormatString = getInstanceFormatString(dbString,
this.txtInstance.Text);

        //SDE 空间连接属性
        IPropertySet propertySet=new PropertySetClass();
        propertySet.SetProperty("server",this.txtServer.Text);
        propertySet.SetProperty("instance",strFormatString);
        propertySet.SetProperty ("database",this.txtDatabase.Text);
        propertySet.SetProperty("user",this.txtUser.Text);
        propertySet.SetProperty ("password",this.txtPassword.Text);
```

```
            //propertySet.SetProperty("version","SDE.DEFAULT");

            //打开 SDE 工作空间
            try
            {
                IWorkspaceFactory workspaceFactory=new SdeWorkspaceFactory
();
                workspace=workspaceFactory.Open(propertySet,0);
                MessageBox.Show("连接 SDE 空间数据库成功");
            }
            catch (Exception ex)
            {
                workspace=null;
                MessageBox.Show("连接 SDE 空间数据库不成功");
            }

        }

        private void btnOK_Click(object sender,EventArgs e)
        {
            this.Close();
        }

        private string getInstanceFormatString(string dbString,string
txtInstance )
        {
            string strFormatString="";
            switch (dbString)
            {
                case "SQL Server":
                    strFormatString="sde:sqlserver:";
                    break;
                case "Oracle":
                    strFormatString="sde:oracle:";
                    break;
                case "DB2":
                    strFormatString="sde:db2:";
                    break;
```

```
            case "PostgreSQL":
                strFormatString="sde:postgresql:";
                break;
        }

        return (strFormatString+txtInstance);
    }
}
```

19.3.2 DbManagerDockFrm 实现

1. DbManagerDockFrm 界面设计

①向左停靠与 TOCControl 窗体叠合在一起，不占用地图显示区域；

②内部采用 TreeView 控件按树状结构显示数据库的内容(类似 ArcCatalog)。

实现效果如图 19-2 所示。

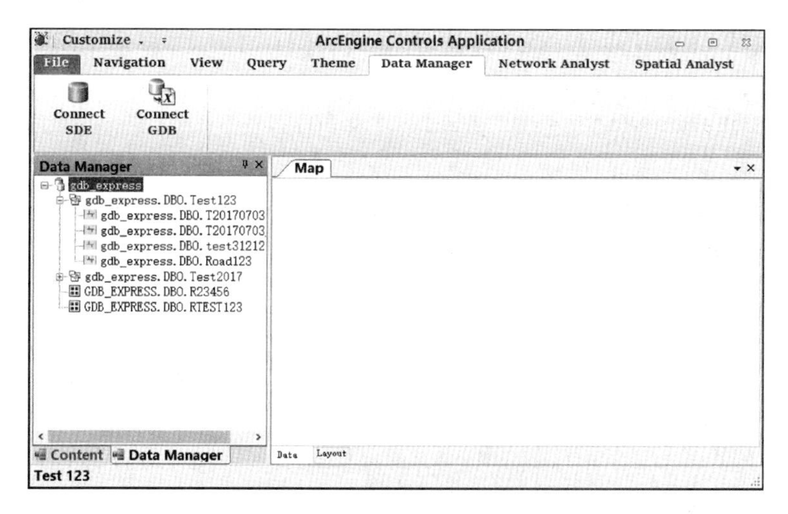

图 19-2　实现效果界面

2. 实现方法

①新建 Windows 窗体，然后将基类改为 DockContent；

②拖入控件 TreeView(名称：treeView1)，设置 dock 属性为 Fill；

③添加私有成员：m_mapControl、m_pWorkspace、m_pTreeNode；

④修改构造函数传入两个参数 IWorkspace、IMapControl3，为私有成员赋值；

⑤装载响应函数，生成树细节结构：

- 以 Database 连接属性生成根节点；
- 遍历数据中的矢量数据集，生成矢量数据集节点元素及其子节点(要素类)；
- 遍历数据中的栅格数据集，生成栅格数据集节点元素。

这里主要用到 IWorkspace 的 get_DatasetNames( )函数获取数据集名称枚举器。

实现代码如下：

```
public partial class DbManagerDockFrm:DockContent
{
    private IMapControl3 m_mapControl=null;
    private IWorkspace m_pWorkspace=null;
    private TreeNode m_pTreeNode=null;
    public DbManagerDockFrm ( IWorkspace pWorkspace, IMapControl3
mapControl)
    {
        InitializeComponent();
        m_pWorkspace=pWorkspace;
        m_mapControl=mapControl;
    }

    private void DbManagerDockFrm_Load(object sender,EventArgs e)
    {
        DatabaseManagerUiInitialize(0);
    }

    private void DatabaseManagerUiInitialize(int iDb)
    {
        IEnumDatasetName enumDatasetName=null;
        IDatasetName datasetName=null;
        if (m_pWorkspace!=null)
        {
            treeView1.Nodes.Clear();
            //添加根节点
            string DatabaseName=
                                m_pWorkspace.ConnectionProperties.GetPro-
perty("database").ToString();
            TreeNode topNode=treeView1.Nodes.Add(DatabaseName);
            topNode.Tag=esriDatasetType.esriDTSchematicDataset;
            topNode.ImageIndex=iDb;
            topNode.SelectedImageIndex=iDb;

            //遍历矢量数据集
            enumDatasetName=
```

```
                    m_pWorkspace.get_DatasetNames(esriData-
setType.esriDTFeatureDataset);
            while ((datasetName=enumDatasetName.Next())!=null)
            {
                TreeNode pNode=topNode.Nodes.Add("FeatureDataset",
datasetName.Name);
                pNode.Tag=esriDatasetType.esriDTFeatureDataset;
                pNode.ImageIndex=2;
                pNode.SelectedImageIndex=2;

                //获取子集名称枚举
                IEnumDatasetName subDsNmae=datasetName.SubsetNames;
                //打开所有子集对象,并加到Map中
                while ((datasetName=subDsNmae.Next())!=null)
                {
                    TreeNode subNode=pNode.Nodes.Add(datasetName.Name);
                    subNode.Tag=esriDatasetType.esriDTFeatureClass;
                    subNode.ImageIndex=6;
                    subNode.SelectedImageIndex=6;
                }
            }
            //遍历栅格数据集
            enumDatasetName=
                    m_pWorkspace.get_DatasetNames(esriData-
setType.esriDTRasterDataset);
            while ((datasetName=enumDatasetName.Next())!=null)
            {
                TreeNode pNode=topNode.Nodes.Add("RasterDataset",data-
setName.Name);
                pNode.Tag=esriDatasetType.esriDTRasterDataset;
                pNode.ImageIndex=3;
                pNode.SelectedImageIndex=3;
            }
        }
    }
}
```

3. 调用

①在RibbonControl上添加Tab项：取名为“Data Manager”，在Tab上创建一个RibbonBar，

再在其上添加两个 ButtonItem 项，分别取名为 ConnectSDE、ConnentGDB。

②在 ConnectSDE 的 Click 响应函数中启动 ConnectSdeServerFrm 窗体。

③如果连接成功("确定")，将 ConnectSdeServerFrm 的 workspace 属性赋给主窗体的 m_pWorkspace成员变量，同时创建 DbManagerDockFrm 停靠在主窗体的左边。

代码如下：

```
private IWorkspace m_pWorkspace=null;
private void btnConnectSDE_Click(object sender,EventArgs e)
{
    ConnectSdeServerFrm frm=new ConnectSdeServerFrm();
    if (frm.ShowDialog()= =System.Windows.Forms.DialogResult.OK)
    {
        m_pWorkspace=frm.workspace;
        DbManagerDockFrm frmDbManager =new DbManagerDockFrm(m _pWork-
space,_mapControl);
        frmDbManager.Show(dockPanel1,DockState.DockLeft);
    }
}
```

19.4 访问 GeoDatabase 数据集

访问 GeoDatabase 数据集可通过扩充 DbManagerDockFrm 实现。

1. 为 DbManagerDockFrm 添加浮动菜单

为 DbManagerDockFrm 添加浮动菜单，并取名为 contextMenuDbManager，添加 Open 菜单项，以及相应的子菜单如表 19-2 所示。

菜单项	子菜单	备　　注
Open	Feature Dataset	打开一个矢量数据集的所要素类
	Feature Class	打开一个要素类
	Raster Dataset	打开一个栅格数据集

2. 子菜单响应函数

相应的子菜单响应函数实现如下：分别调用 OpenOpsClass(打开操作)的方法。

```
//打开一个要素类
private void openfeatureClassToolStripMenuItem_Click(object sender,
EventArgs e)
{
    esriDatasetType dsType=(esriDatasetType)m_pTreeNode.Tag;
```

```
        //限于要素数据类节点
        if (dsType==esriDatasetType.esriDTFeatureClass)
        {
            string dsName=m_pTreeNode.FullPath;
            int lastIndex=dsName.LastIndexOf(@"\");
            dsName=dsName.Substring(lastIndex+1);

            OpenOpsClass Ops=new OpenOpsClass(m_pWorkspace,m_mapControl);
            Ops.OpenFeatureClass(dsName);
        }
    }
    //打开一个矢量数据集的所有要素类
    private void openfeatuerDatasetToolStripMenuItem_Click(object sender,EventArgs e)
    {
        esriDatasetType dsType=(esriDatasetType)m_pTreeNode.Tag;
        //限于要素数据集节点
        if (dsType==esriDatasetType.esriDTFeatureDataset)
        {
            string dsName=m_pTreeNode.FullPath;
            int lastIndex=dsName.LastIndexOf(@"\");
            dsName=dsName.Substring(lastIndex+1);

            OpenOpsClass Ops=new OpenOpsClass(m_pWorkspace,m_mapControl);
            Ops.OpenFeatureDataset(dsName);
        }
    }
    //打开一个栅格数据集
    private void openRasterDatasetToolStripMenuItem1_Click(object sender,EventArgs e)
    {
        esriDatasetType dsType=(esriDatasetType)m_pTreeNode.Tag;
        //限于栅格数据集节点
        if (dsType==esriDatasetType.esriDTRasterDataset)
        {
            string dsName=m_pTreeNode.FullPath;
            int lastIndex=dsName.LastIndexOf(@"\");
            dsName=dsName.Substring(lastIndex+1);
```

```
            OpenOpsClass Ops =new OpenOpsClass(m_pWorkspace,m_mapControl);
            Ops.OpenRasterDataset(dsName);
        }
    }
```

3. 激活 contextMenuDbManager 菜单

通过 TreeView 节点右键响应函数激活 contextMenuDbManager 菜单，同时记录鼠标点击的节点对象，代码如下：

```
    private void treeView1_NodeMouseClick(object sender,TreeNodeMouse-
ClickEventArgs e)
    {
        if (e.Button = =MouseButtons.Right)
        {
            m_pTreeNode=treeView1.GetNodeAt(e.X,e.Y);
            contextMenuDbManager.Show(this.treeView1,e.X,e.Y);
        }
    }
```

19.5 OpenOpsClass 功能类实现

打开矢量数据集的步骤如下：

①用 IWorkspace 的 get_DatasetNames()函数获取数据集名称枚举器(IDatasetName)。

②遍历 IDatasetName 的子类型：

- 获取子类型要素类的名称；
- 使用 IFeatureWorkspace 的 OpenFeatureClass()函数打开 IFeatureClass；
- 将 IFeatureClass 添加到 Map 图层中；
- 重复此过程直至所有子类型结束。

③如果已知要素类名，则可直接用 OpenFeatureClass()打开。

④栅格数据集使用 IRasterWorkspaceEx 接口的 OpenRasterDataset()打开。

浏览数据功能封装在 OpenOpsClass 类中，代码如下：

```
public class OpenOpsClass
{
    private IWorkspace m_pWorkspace=null; //工作空间
    private IMapControl3 m_mapControl=null;
    public OpenOpsClass(IWorkspace workspace,IMapControl3 mapControl)
    {
        m_pWorkspace=workspace;
        m_mapControl=  mapControl;
```

```
        }

        // 加载矢量数据集
        public void OpenFeatureDataset(string dsName)
        {
            //获取数据集名称对象
            IDatasetName datasetName=QueryDatasetByname(m_pWorkspace,
                                                        esriDatasetType. esri-
DTFeatureDataset,dsName);
            if (datasetName==null)
                return;

            //获取子集名称枚举
            IEnumDatasetName subDsNmae=datasetName.SubsetNames;

            //打开所有子集对象,并加到 Map 中
            while ((datasetName=subDsNmae.Next())!=null)
            {
                IFeatureWorkspace featureWorkspace=m_pWorkspace as IF-
eatureWorkspace;
                IFeatureClass sdeFeatureClass =
                                                featureWorkspace.OpenFea-
tureClass(datasetName.Name);

                //加载数据到 Mapcontrol
                IFeatureLayer sdeFeatureLayer=new FeatureLayerClass();
                sdeFeatureLayer.FeatureClass=sdeFeatureClass;
                sdeFeatureLayer.Name=datasetName.Name;
                m_mapControl.AddLayer(sdeFeatureLayer as ILayer,0);
                m_mapControl.Extent=this.m_mapControl.FullExtent;
            }
        }

        //加载要素类
        public void OpenFeatureClass(string className)
        {
            IFeatureWorkspace featureWorkspace=m_pWorkspace as IFeat-
ureWorkspace;
```

```
            IFeatureClass sdeFeatureClass = featureWorkspace.OpenFeature-
Class(className);

            //加载数据到 Mapcontrol
            IFeatureLayer sdeFeatureLayer=new FeatureLayerClass();
            sdeFeatureLayer.FeatureClass=sdeFeatureClass;
            sdeFeatureLayer.Name=className;
            m_mapControl.AddLayer(sdeFeatureLayer as ILayer,0);
            m_mapControl.Extent=this.m_mapControl.FullExtent;
        }

        //加载影像数据
        public void OpenRasterDataset(string dsName)
        {
            //获取数据集名称对象
            IDatasetName datasetName=QueryDatasetByname(m_pWorkspace,
                                                        esriDatasetType. es-
riDTRasterDataset,dsName);
            if (datasetName==null)
                return;

            //打开栅格数据集
            IRasterWorkspaceEx rasterWorksapce=m_pWorkspace as IRaste-
rWorkspaceEx;
            IRasterDataset rasterDataset=rasterWorksapce.OpenRasterDataset
(datasetName.Name);

            //加载数据到 Mapcontrol
            IRasterLayer rasterLayer=new RasterLayerClass ();
            rasterLayer.CreateFromDataset(rasterDataset);
            m_mapControl.AddLayer(rasterLayer as ILayer,0);
            m_mapControl.Extent=m_mapControl.FullExtent;
        }
    }
```

用到辅助函数：QueryDatasetByname，代码如下：

```
/// <summary>
///在工作空间中查询指定名称的数据集
/// </summary>
```

```
    /// <param name="pWorkspace"></param>
    /// <param name="dsName"></param>
    /// <returns></returns>
    private IDatasetName QueryDatasetByname(IWorkspace pWorkspace,es-
riDatasetType dsType,string dsName)
    {
        if (pWorkspace==null)
            return null;

        //获取矢量数据集名称对象
        IEnumDatasetName enumDatasetName=pWorkspace.get_DatasetNames
(dsType);

        IDatasetName datasetName=null;
        bool isExist=false;
        while ((datasetName=enumDatasetName.Next())!=null)
        {
            if (datasetName.Name==dsName)
            {
                isExist=true;
                break;
            }
        }

        return (isExist==true) ? datasetName:null;
    }
```

第 20 章　空间数据建库

20.1　概述

本章在上一章的基础上，扩展空间数据建库功能，主要包括：创建数据集(矢量数据集、栅格数据集)，创建要素类，数据入库(矢量数据集、栅格数据集)。

使用环境：

①测试数据库：Microsoft SQL Server 2016 Express；

②操作系统：Windows 7 SP1(64 位)，机器名为 WIN-KG9LKA8CBST，注意关闭系统防火墙。

20.2　数据库存储结构

20.2.1　CreateFeatureClassFrm 功能类

1. 功能描述

CreateFeatureClassFrm 实现新建要素类功能，难点是创建要素字段集。本功能类设计为一个 Form 窗体，能提供字段集编辑表单界面(采用 DataGridView)，也可根据 shp 模板文件克隆一个字段集。界面设计如图 20-1 所示。

图 20-1　界面设计

2. 功能实现

(1)新建功能窗体

项目中添加一个新的窗体，名称为"CreateFeatureClassFrm"，拖动 DataGridView、ComBox、TextBox、Button 等控件到窗体，见表 20-1。

表 20-1　**控件属性设置**

控件类型	Name 属性	控件说明	备　注
TextBox	txtFeatureClassName	要素类名称	
ComBox	cbxShpType	几何类型	
DataGridView	dataGridView1	字段集合列表	第二列(数据类型)设计为 DataGridViewComboBoxColumn
Button	btnExplor	浏览模板文件名	
Button	btnOK	确定	
Button	btnApp	应用	
Button	btnCancel	取消	

cbxShpTypex 和 dataGridView1 的 DataGridViewComboBoxColumn 列的可选值，参考 StringToFieldType( )和 StringToGeometryType( )在设计时填充。

(2)类设计

代码如下：

```
public partial class CreateFeatureClassFrm:Form
{
    private IWorkspace m_pWorkspace=null; //工作空间
    private string m_NameOfDataset="";
    private string m_NameOfTemplate="";
    private string m_NameOfFeatureClass="";

    public string _NameOfFeatureClass
    {
        get { return m_NameOfFeatureClass;  }
    }
    public CreateFeatureClassFrm(IWorkspace workspace,string dsName)
    {
        InitializeComponent();
        m_pWorkspace=workspace;
        m_NameOfDataset=dsName;
```

```
    }
    //消息响应函数
    private void CreateFeatureFrm_Load(object sender,EventArgs e)
    private void btnExplor_Click(object sender,EventArgs e)
    private void btnOK_Click(object sender,EventArgs e)
    private void btnApp_Click(object sender,EventArgs e)

    //核心功能函数
    private IFields CloneFeatureClassFields(IFeatureClass pfc,ISpatialReference spatialReference2)
    private IFields CreateFeatureClassFields(esriGeometryType shapeType,ISpatialReference spatialReference2)
    private IFeatureClass CreateFeatureClass(esriGeometryType shapeType,string targetDsName,string TargetFCname)

    //辅助函数
    private IGeometryDef CreateGeometryDef(esriGeometryType shapeType,ISpatialReference spatialReference2)
    private IFeatureClass OpenFeatureClassByShpfile (string shpFilePath,string shpFileName)
    ......
}
```

(3)响应函数实现

代码如下:

```
//装载响应函数,在 DataGridView 中填充两个必需字段
private void CreateFeatureFrm_Load(object sender,EventArgs e)
{
    int index=this.dataGridView1.Rows.Add();
    DataGridViewRow row=this.dataGridView1.Rows[index];
    row.Cells[0].Value="OBJECTID";
    row.Cells[1].Value="OID";

    index=this.dataGridView1.Rows.Add();
    row=this.dataGridView1.Rows[index];
    row.Cells[0].Value="shape";
    row.Cells[1].Value="Geometry";
}
//浏览模板数据文件,填充 DataGridView
```

```
private void btnExplor_Click(object sender,EventArgs e)
{
    this.openFileDialog1.Filter="shp file (*.shp) |*.shp";
    this.openFileDialog1.Title="打开矢量数据";
    this.openFileDialog1.Multiselect=false;
    if (this.openFileDialog1.ShowDialog()= =DialogResult.OK)
    {
        m_NameOfTemplate=this.openFileDialog1.FileName;
        int lastIndex=m_NameOfTemplate.LastIndexOf(@ "\");
        string shpFilePath=m_NameOfTemplate.Substring(0,lastIndex);
        string shpFileName=m_NameOfTemplate.Substring(lastIndex+1);

        //打开指定 ShapeFile 要素类
        this.dataGridView1.Rows.Clear();
        IFeatureClass shpfc =OpenFeatureClassByShpfile(shpFilePath,shp-
FileName);
        for (int i=0; i < shpfc.Fields.FieldCount; i++)
        {
            IField pFd=shpfc.Fields.get_Field(i);
            int index=this.dataGridView1.Rows.Add();
            DataGridViewRow row=this.dataGridView1.Rows[index];
            row.Cells[0].Value=pFd.Name;
            row.Cells[1].Value=pFd.Type.ToString().Substring(13);
        }
    }
}

//OK
private void btnOK_Click(object sender,EventArgs e)
{
    btnApp_Click(sender,e);
    this.Close();
}
//APP 函数调 CreateFeatureClass
private void btnApp_Click(object sender,EventArgs e)
{
    if (this.cbxShpType.SelectedIndex < 0 ||this.txtFeatureClassName.Text =
="")
```

```
            return;

        string shpTypeString=this.cbxShpType.SelectedItem.ToString();
        esriGeometryType shapeType=StringToGeometryType( shpTypeString );
        m_NameOfFeatureClass=this.txtFeatureClassName.Text;

        CreateFeatureClass(shapeType,m_NameOfDataset,m_NameOfFeature-
Class);
    }
```

(4)核心函数实现

1)字段集创建函数：CreateFeatureClassFields( )

要素字段集必须包括 OID(字段名 OBJECTID)、Geometry(字段名通常是 shape)两个特殊类型的字段，分别表示对象 ID 和几何图形，其他字段为用户自定义字段，一般只要指定字段名和字段数据类型。主要用到 FieldClass、FieldsClass 以及相关接口。

FieldClass 实现 IField、IFieldEdit 接口：

①IField 为字段基本接口；

②IFieldEdit 为字段编辑接口，可为字段设置名称、数据类型等。

③对于 shape 字段还需要配置 IGeometryDefEdit 接口属性，该字段包含几何类型、空间参考系、空间索引等内容。

FieldsClass 实现 IFields 字段集接口，是 IField 的集合。

这里 shape 字段此函数用到 CreateGeometryDef( )辅助函数创建 IGeometryDef 对象，定义几何信息存储字段的空间索引和空间参考系，是创建要素类的要点，具体代码如下：

```
    private IFields CreateFeatureClassFields(esriGeometryType shape-
Type,ISpatialReference spatialReference2)
    {
        //创建新的字段集
        IFields pFields=new FieldsClass();
        IFieldsEdit pFieldsEdit=(IFieldsEdit)pFields;

        //遍历 DataGridView 行,每行建立一个字段
        for (int i=0; i < this.dataGridView1.Rows.Count-1; i++)
        {
            string NameOfString=this.dataGridView1[0,i].Value.ToString();
            string TypeOfString=this.dataGridView1[1,i].Value.ToString();
            switch (i)
            {
                case 0://产生新的 FID 字段
                    {
```

```
                    IField pField=new FieldClass();
                    IFieldEdit pFieldEdit=(IFieldEdit)pField;
                    pFieldEdit.Name_2="OBJECTID";
                    pFieldEdit.AliasName_2="OBJECTID";
                    pFieldEdit.Type_2=esriFieldType.esriFieldTypeOID;
                    pFieldsEdit.AddField(pField);
                }
                break;
            case 1://产生新的 shape 字段
                {
                    IField pField=new FieldClass();
                    IFieldEdit pFieldEdit=(IFieldEdit)pField;
                    pFieldEdit.Name_2="shape";
                    pFieldEdit.AliasName_2="shape";
                    pFieldEdit.Type_2=esriFieldType.esriFieldTypeGeome-
try;
                    pFieldEdit.GeometryDef_2=CreateGeometryDef(shape-
Type,spatialReference2);
                    pFieldsEdit.AddField(pField);
                }
                break;
            default://产生自定义字段
                {
                    IField pField=new FieldClass();
                    IFieldEdit pFieldEdit=(IFieldEdit)pField;
                    pFieldEdit.Name_2=NameOfString;
                    pFieldEdit.AliasName_2=NameOfString;
                    pFieldEdit.Type_2=StringToFieldType(TypeOfString);
                    pFieldsEdit.AddField(pField);
                }
                break;
        }
    }

    return pFields;
}
```

2)要素类创建函数：CreateFeatureClass()

ArcGIS 数据库对矢量数据按两级组织，第一级是矢量数据集集合(简称数据集)，第二级包含若干要素类。矢量数据集创建成功后，可在其下创建要素类。步骤如下：

①创建要素字段集，可新建或可根据 shp 文件克隆一个字段集；

②建立一个要素对象描述(可以没有)；

③打开数据集，用 IFeatureWorkspace 的 OpenFeatureDataset()；

④创建要素数据类，用数据集 IFeatuerDataset 的 CreateFeatureClass()方法。

源代码如下：

```
///创建要素类
private IFeatureClass CreateFeatureClass(esriGeometryType shapeType,
                                         string targetDsName,string
TargetFCname)
{
    //判断是否在 SDE 中已经存在要素类
    IWorkspace2 pW2 =m_pWorkspace as IWorkspace2;
    if (!pW2.get_NameExists(esriDatasetType.esriDTFeatureClass,
TargetFCname))
    {
        //打开指定数据集
        IFeatureWorkspace featureWorkspace =m_pWorkspace as IFeat-
ureWorkspace;
        IFeatureDataset fDataset =featureWorkspace.OpenFeatureDataset
(targetDsName);

        //创建要素类的字段集
        ISpatialReference pSpatialReference =(fDataset as IGeoData-
set).SpatialReference;
        IFields pFields =CreateFeatureClassFields(shapeType,pSpa-
tialReference);

        //要素对象描述
        IFeatureClassDescription fcDescript =new FeatureClassDescrip-
tionClass();
        IObjectClassDescription ObjDescript =fcDescript as IObject-
ClassDescription;

        //在数据库中创建矢量数据层
        return fDataset.CreateFeatureClass(TargetFCname,pFields,
ObjDescript.Inst-anceCLSID,ObjDescript.ClassExtensionCLSID,esriFea-
tureType.esriFTSimple,"shape","");
```

```
        }

        return null;
    }
```

(5)辅助函数

代码如下：

```
    ///空间参考与空间索引
    private IGeometryDef CreateGeometryDef(esriGeometryType shapeType,
                                           ISpatialReference spatial-
Reference2)
    {
        IGeometryDef geometryDef=new GeometryDefClass();
        IGeometryDefEdit geometryDefedit=(IGeometryDefEdit)geometryDef;
        //平均点数
        geometryDefedit.AvgNumPoints_2=5;
        //空间索引
        geometryDefedit.GridCount_2=1;
        geometryDefedit.set_GridSize(0,100000);
        //空间数据类型
        geometryDefedit.GeometryType_2=shapeType;
        //空间参考系
        geometryDefedit.SpatialReference_2=spatialReference2;
        return geometryDef
    }

    private esriGeometryType StringToGeometryType(string TypeOfString)
    {
        switch (TypeOfString)
        {
            case "Point":
                return esriGeometryType.esriGeometryPoint;
                break;
            case "Polyline":
                return esriGeometryType.esriGeometryPolyline;
                break;
            case "Polygon":
                return esriGeometryType.esriGeometryPolygon;
                break;
```

```
            case "Multipoint":
                return esriGeometryType.esriGeometryMultipoint;
                break;
            case "MultiPatch":
                return esriGeometryType.esriGeometryMultiPatch;
                break;
            default:
                return esriGeometryType.esriGeometryPoint;
                break;
        }

    }
    private esriFieldType StringToFieldType(string TypeOfString)
    {
        switch (TypeOfString)
        {
            case "OID":
                return esriFieldType.esriFieldTypeOID;
                break;
            case "Geometry":
                return esriFieldType.esriFieldTypeGeometry;
                break;
            case "GUID":
                return esriFieldType.esriFieldTypeGUID;
                break;
            case "String":
                return esriFieldType.esriFieldTypeString;
                break;
            case "SmallInteger":
                return esriFieldType.esriFieldTypeSmallInteger;
                break;
            case "Integer":
                return esriFieldType.esriFieldTypeInteger;
                break;
            case "Single":
                return esriFieldType.esriFieldTypeSingle;
                break;
            case "Double":
```

```
                return esriFieldType.esriFieldTypeDouble;
                break;
            case "Date":
                return esriFieldType.esriFieldTypeDate;
                break;
            case "Blob":
                return esriFieldType.esriFieldTypeBlob;
                break;
            case "XML":
                return esriFieldType.esriFieldTypeXML;
                break;
            default:
                return esriFieldType.esriFieldTypeString;
                break;
        }
    }
    private IFeatureClass OpenFeatureClassByShpfile(string shpFilePath,
string shpFileName)
    {
        //打开 SHP 数据
        IWorkspaceFactory shpwpf=new ShapefileWorkspaceFactoryClass();
        IFeatureWorkspace shpfwp = shpwpf.OpenFromFile(shpFilePath,0)
as IFeatureWorkspace;
        IFeatureClass shpfc=shpfwp.OpenFeatureClass(shpFileName);

        return shpfc;
    }
```

20.2.2　CreateDatasetFrm 功能类

1. 功能描述

本类设计为同时支持矢量数据集和栅格数据集，具体情况依据构造函数的 esriDatasetType 参数决定，有关参数从 DataGridView 获取，用户界面如图 20-2 所示。

2. 功能实现

(1)类设计

具体代码如下：

```
public partial class CreateDatasetFrm:Form
{
    private IWorkspace m_pWorkspace=null;
```

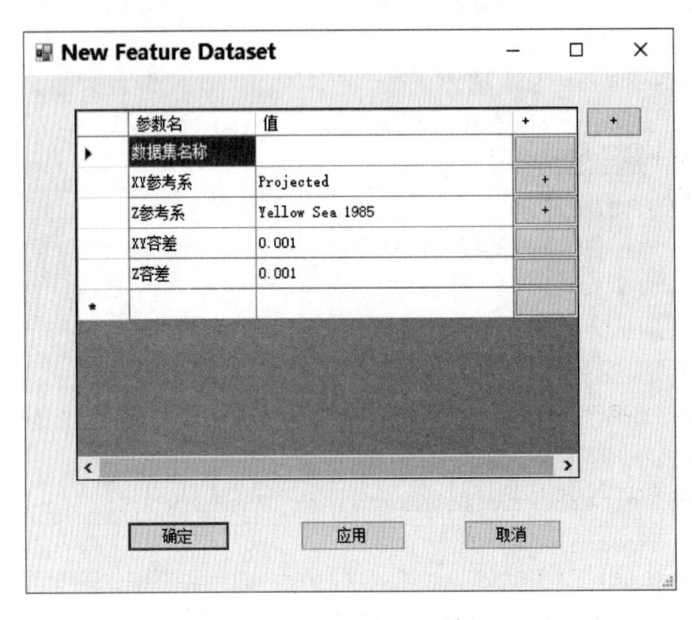

图 20-2　New Feature Dataset 界面

```
private esriDatasetType m_datasetType = esriDatasetType.esriDT-
Any;
private string m_NameOfDataset = "";
public string _NameOfDataset
{
    get { return m_NameOfDataset; }
}
//构造函数
public CreateDatasetFrm(IWorkspace workspace, esriDatasetType
dsType)
//消息响应函数
private void CreateFeatureDatasetFrm_Load(object sender, Even-
tArgs e)
private void btnBrowse_Click(object sender, EventArgs e)
private void btnApp_Click(object sender, EventArgs e)
private void btnOK_Click(object sender, EventArgs e)
//核心函数
private void CreateFDataset()
private void CreateRDataset()
//辅助函数
private ISpatialReference CreateSpatialReference(string xyRef-
```

```
erence,string zReference,double xyTolerance,double zTolerance,IEnvelope
pDomainEnv)
        ......
    }
```

(2)构造函数+消息响应函数

具体代码如下：

```
    public CreateDatasetFrm ( IWorkspace workspace, esriDatasetType
dsType )
    {
        InitializeComponent();
        m_pWorkspace=workspace;
        m_datasetType=dsType;
        switch (dsType)
        {
            case esriDatasetType.esriDTFeatureDataset:
                this.Text="New Feature Dataset";
                break;
            case esriDatasetType.esriDTRasterDataset:
                this.Text="New Raster Dataset";
                break;
        }
    }

    private void CreateFeatureDatasetFrm_Load(object sender,EventArgs e)
    {
        switch (m_datasetType)
        {
            case esriDatasetType.esriDTFeatureDataset:
                {
                    NewDataGridViewRow("数据集名称","","");
                    NewDataGridViewRow("XY 参考系","Projected","+");
                    NewDataGridViewRow("Z 参考系","Yellow Sea 1985","+");
                    NewDataGridViewRow("XY 容差","0.001","");
                    NewDataGridViewRow("Z 容差","0.001","");
                }
                break;
            case esriDatasetType.esriDTRasterDataset:
                {
```

```
                    NewDataGridViewRow("数据集名称","","");
                    NewDataGridViewRow("XY 参考系","Projected","+");
                    NewDataGridViewRow("Z 参考系","Yellow Sea 1985","+");
                    NewDataGridViewRow("分辨率","100,100","");
                    NewDataGridViewRow("像素类型","9","");
                }
                break;
        }
    }

    private void NewDataGridViewRow(string NameOfParameter,string Val-
ueOfParameter,string Extension)
    {
        int index=this.dataGridView1.Rows.Add();
        DataGridViewRow row=this.dataGridView1.Rows[index];
        row.Cells[0].Value=NameOfParameter;
        row.Cells[1].Value=ValueOfParameter;
        row.Cells[2].Value=Extension;
    }
    //获取模板栅格数据参数
    private void btnBrowse_Click(object sender,EventArgs e)
    {
        this.openFileDialog1.Filter="tif file (*.tif) |*.* ";
        this.openFileDialog1.Title="打开栅格数据";
        this.openFileDialog1.Multiselect=false;
        if (this.openFileDialog1.ShowDialog()= =DialogResult.OK)
        {
            string NameOfTemplate=this.openFileDialog1.FileName;
            int lastIndex=NameOfTemplate.LastIndexOf(@ " \");
            string FilePath=NameOfTemplate.Substring(0,lastIndex);
            string FileName=NameOfTemplate.Substring(lastIndex+1);

            //打开栅格数据
            IWorkspaceFactory tifwpf=new RasterWorkspaceFactoryClass();
            IWorkspace tifwp=tifwpf.OpenFromFile(FilePath,0);
            IRasterWorkspace tifrwp=tifwp as IRasterWorkspace;
            IRasterDataset ipRasterDataset = tifrwp.OpenRasterDataset(File-
Name);
```

```
            IRaster ipRaster=ipRasterDataset.CreateDefaultRaster();

            IRasterProps rasterProps=(IRasterProps)ipRaster;
            IPnt pt=rasterProps.MeanCellSize();
            string sSize=pt.X.ToString()+","+pt.Y.ToString();
            string sPixel=((int)rasterProps.PixelType).ToString();

            this.dataGridView1.Rows.Clear();
            NewDataGridViewRow("数据集名称","","");
            NewDataGridViewRow("XY 参考系",rasterProps.SpatialReference.
Name,"+");
            NewDataGridViewRow("Z 参考系","Yellow Sea 1985","+");
            NewDataGridViewRow("分辨率",sSize,"");
            NewDataGridViewRow("像素类型",sPixel,"");
        }
    }

    private void btnApp_Click(object sender,EventArgs e)
    {
        switch (m_datasetType)
        {
            case .esriDTFeatureDataset:
                CreateFDataset();
                break;
            case esriDatasetType.esriDTRasterDataset:
                CreateRDataset();
                break;
        }
    }

    private void btnCancel_Click(object sender,EventArgs e)
    {
        this.Close();
    }
    //创建矢量数据集
    private void btnOK_Click(object sender,EventArgs e)
    {
        btnApp_Click( sender,  e);
```

```
        this.Close();
    }
```

(3)核心函数

1)创建矢量数据集函数：CreateFDataset()

矢量数据集用 IFeatureWorkspace 的 CreateFeatureDataset()函数创建。需要以下参数：数据集名(dsName)、空间参考系(spatialReference)、容差(Tolerance)。

代码如下：

```
    private void CreateFDataset()
    {
        //工作空间查找指定名称数据集
        m_NameOfDataset=this.dataGridView1[1,0].Value.ToString();
        //string dsNameEx="DBO."+this.dataGridView1[1,0].Value.ToS-
tring();
        string xyReference=this.dataGridView1[1,1].Value.ToString();
        string zReference=this.dataGridView1[1,2].Value.ToString();
        double xyTolerance = double.Parse(this.dataGridView1[1,3].
Value.ToString());
        double zTolerance = double.Parse(this.dataGridView1[1,4].
Value.ToString());
        //设置范围矩形
        IEnvelope pDomainEnv=new EnvelopeClass();
        pDomainEnv.XMin=-99999999; pDomainEnv.XMax=99999999;
        pDomainEnv.YMin=-99999999; pDomainEnv.YMax=99999999;

        IDatasetName datasetName=QueryDatasetByname(m_pWorkspace,
                                    esriDatasetType.esriDTFeatureData-
set,m_NameOfDataset);
        if (datasetName!=null)
        {
            MessageBox.Show("数据集已存在");
            return;
        }
        try
        {
            //定义空间参考
             ISpatialReference sr=CreateSpatialReference(xyReference,
zReference,xyTolerance,zTolerance,pDomainEnv);
            //创建矢量数据集
```

```
            IFeatureWorkspace ftWorkspace=m_pWorkspace as IFeatureWork-
space;
            IFeatureDataset fDataset=ftWorkspace.CreateFeatureDataset
(m_NameOfDataset,sr);
        }
        catch (Exception ex)
        {
            MessageBox.Show(ex.ToString()+"+++"+ex.Message);
        }
    }
```

2)创建栅格数据集函数：CreateRDataset( )

创建栅格数据集函数的步骤如下：

①设置存储结构：IRasterStorageDef；

②设置栅格数据集空间坐标系：IRasterDef；

③设置栅格数据集空间索引：IGeometryDef；

④创建栅格数据集：IRasterWorkspaceEx 的 CreateRasterDataset( )函数。

代码如下：

```
    private void CreateRDataset()
    {
        m_NameOfDataset=this.dataGridView1[1,0].Value.ToString();
        string xyReference=this.dataGridView1[1,1].Value.ToString();
        string zReference=this.dataGridView1[1,2].Value.ToString();
        string[] sizeArr=this.dataGridView1[1,3].Value.ToString().
Split(new char[] {'  ,' });
        rstPixelType pixelType=(rstPixelType) int.Parse(this.data-
GridView1[1,4].Value.ToString());

        //设置范围矩形
        IEnvelope pDomainEnv=new EnvelopeClass();
        pDomainEnv.XMin=-99999999; pDomainEnv.XMax=99999999;
        pDomainEnv.YMin=-99999999; pDomainEnv.YMax=99999999;

        IDatasetName datasetName=QueryDatasetByname(m_pWorkspace,
                                    esriDatasetType. esriDTRasterData-
set,m_NameOfDataset);
        if (datasetName!=null)
        {
            MessageBox.Show("数据集已存在");     return;
```

```
    }

    try
    {
        //设置存储参数
        IRasterStorageDef rasterStorageDef = new RasterStorageDef-
Class();
        rasterStorageDef.CompressionType =
                        esriRasterCompressionType. esriRasterCom-
pressionUncompressed;
        rasterStorageDef.PyramidLevel = 1;
        rasterStorageDef.PyramidResampleType = rstResamplingTypes.
RSP_NearestNeighbor;
        rasterStorageDef.TileHeight = 128;
        rasterStorageDef.TileWidth = 128;

        IPnt pt = new Pnt();
        pt.SetCoords(double.Parse(sizeArr[0]), double.Parse(sizeArr
[1]));
        rasterStorageDef.CellSize = pt;

        //设置栅格数据集空间坐标系
        IRasterDef rasterDef = new RasterDefClass();
        //rasterDef.SpatialReference = new UnknownCoordinateSystem-
Class();
        rasterDef.SpatialReference = CreateSpatialReference(xyRef-
erence,zReference,0.001,0.001,pDomainEnv);
        //设置栅格数据集空间索引
        IGeometryDef geometryDef = new GeometryDefClass();
        IGeometryDefEdit geometryDefedit = (IGeometryDefEdit)geome-
tryDef;
        geometryDefedit.AvgNumPoints_2 = 5;
        geometryDefedit.GridCount_2 = 1;
        geometryDefedit.set_GridSize(0,1000);
        geometryDefedit.GeometryType_2 = esriGeometryType.esriGeom-
etryPolygon;
        geometryDefedit.SpatialReference_2 = CreateSpatialReference
(xyReference,zReference,
```

```
 0.001,0.001,pDomainEnv);
            //创建栅格数据集
            IRasterWorkspaceEx rWorkspace=m_pWorkspace as IRasterWork-
spaceEx;
            IRasterDataset rDataset=rWorkspace.CreateRasterDataset(m_
NameOfDataset,1,
                                        pixelType,rasterStorageDef,"DEFAULTS",
rasterDef,geometryDef);
        }
        catch(Exception ex)
        {
            MessageBox.Show(ex.ToString()+"+++"+ex.Message);
        }
    }
```

(4)辅助函数

注意：这里坐标系的选择做了简化处理，仅提供两种选择——Geographic 和 Projected。

```
    private ISpatialReference CreateSpatialReference(string xyRefer-
ence,string zReference,double xyTolerance,double zTolerance,IEnvelope
pDomainEnv)
    {
        //初始化空间参考系
        ISpatialReferenceFactory spatialReferenceFactory=new Spatial-
ReferenceEnvironmentClass();
        ISpatialReference spatialReference=null;
        switch (xyReference)
        {
            case "Geographic":
                  spatialReference = spatialReferenceFactory.CreateGeo-
graphicCoordinateSystem(
                                                      (int) esriSRGeoCS-
Type.esriSRGeoCS_Beijing1954);
                break;
            case "Projected":
                  spatialReference = spatialReferenceFactory.CreatePro-
jectedCoordinateSystem(
                              (int) esriSRProjCS4Type. esriSRProjCS _ Bei-
jing1954_3_Degree_GK_Zone_40);
                break;
```

```
            default:
                spatialReference=new UnknownCoordinateSystemClass();
                break;
        }

        //设置容差
        ISpatialReferenceTolerance srTolerance=spatialReference as IS-
patialReferenceTolerance;
        srTolerance.XYTolerance=xyTolerance;
        srTolerance.ZTolerance=zTolerance;

        //设置分辨率
        ISpatialReferenceResolution srResolution = spatialReference as
ISpatialReferenceResolution;
        srResolution.ConstructFromHorizon();
        srResolution.SetDefaultXYResolution();

        //设置范围
        spatialReference.SetDomain(pDomainEnv.XMin,pDomainEnv.XMax,
                                  pDomainEnv.YMin,pDomainEnv.YMax);
        return spatialReference;
    }

    //在工作空间中查询指定名称的数据集
    private IDatasetName QueryDatasetByname(IWorkspace pWorkspace,es-
riDatasetType dsType,string dsName)
    {
        if (pWorkspace==null)
            return null;

        //获取矢量数据集名称对象
        IEnumDatasetName enumDatasetName=pWorkspace.get_DatasetNames
(dsType);

        IDatasetName datasetName=null;
        bool isExist=false;
        while ((datasetName=enumDatasetName.Next())!=null)
        {
```

```
        if (datasetName.Name.IndexOf(dsName)>0)
        //if (datasetName.Name==dsName)
        {
            isExist=true;
            break;
        }
    }

    return (isExist==true) ? datasetName:null;
}
```

20.3 数据入库功能实现

LoadOpsClass 函数可实现数据入库的功能。代码如下：

```
class LoadOpsClass
{
    private IWorkspace m_pWorkspace=null;
    public LoadOpsClass(IWorkspace workspace)
    {
        m_pWorkspace=workspace;
    }

    //矢量数据入库
    public void LoadShapefileToDatabase(string shpFilePath,string
shpFileName,string targetFCname)
    //影像图入库
    public bool LoadRasterIntoDbDataset(string fileFullName,string
strDbRasterDatasetName)
    //辅助函数
    private IRasterDataset OpenRasterFromFile(string fileName)
}
```

1. 矢量数据入库函数：LoadShapefileToDatabase()

矢量数据入库的步骤如下：

①打开待入库的 shp 文件；

②打开 SDE 数据库中目标要素类；

③分别建立要素类的访问游标：对 shp 要素类建立 Search 游标，对 GDB 要素类建立 Insert 游标；

④将 shp 要素类中的要素逐个添加到 SDE 要素类中。方法是：先构造一个缓存要素

IFeatureBuffer，用 shp 要素字段值对缓存对象进行赋值，最后用 SDE 游标的 InsertFeature()方法，将缓存对象插入到目标数据集。

⑤每插入 1000 个推送一次。

源代码如下：

```
public void LoadShapefileToDatabase(string shpFilePath,string shpFileName,string targetFCname)
{
    //打开 shp 数据
    IWorkspaceFactory shpwpf=new ShapefileWorkspaceFactoryClass();
    IFeatureWorkspace shpfwps = shpwpf.OpenFromFile(shpFilePath,0) as IFeatureWorkspace;
    IFeatureClass shpfc = shpfwps.OpenFeatureClass(shpFileName);

    IWorkspace2 pW2 =m_pWorkspace as IWorkspace2;
    if (pW2.get_NameExists(esriDatasetType.esriDTFeatureClass,targetFCname))
    {
        //在 SDE 数据库中打开矢量数据层
        IFeatureWorkspace featureWorkspace = m_pWorkspace as IFeatureWorkspace;
        IFeatureClass sdeFeatureClass = featureWorkspace.OpenFeatureClass(targetFCname);

        //分别建立访问要素类的游标
        IFeatureCursor featureCursor=shpfc.Search(null,true);
        IFeatureCursor sdeFeatureCursor=sdeFeatureClass.Insert(true);

        //逐个添加实体对象到数据库
        int numuric=0;
        IFeature feature=default(IFeature);
        IFeatureBuffer sdeFeatureBuffer;
        while ((feature=featureCursor.NextFeature())!=null)
        {
            sdeFeatureBuffer=sdeFeatureClass.CreateFeatureBuffer();
            IField shpField=new FieldClass();
            IFields shpFields=feature.Fields;
            for (int i=0; i < shpFields.FieldCount; i++)
            {
```

```
                    shpField=shpFields.get_Field(i);
                    if (shpField.Type==esriFieldType.esriFieldTypeOID)
                        continue;

                    int index=sdeFeatureBuffer.Fields.FindField(shpField.
Name);
                    if (index!=-1)
                    {
                            sdeFeatureBuffer.set_Value(index, feature.get_
Value(i));
                    }
                }
                sdeFeatureCursor.InsertFeature(sdeFeatureBuffer);

                //每插入 1000 个要素,做一次推送;
                numuric++;
                if (numuric > 1000)
                {
                    sdeFeatureCursor.Flush();
                    numuric=0;
                }
            }

                sdeFeatureCursor.Flush();
        }
    }
```

2. 栅格数据入库：LoadRasterIntoDbDataset()

栅格数据入库的步骤如下：

①打开 SDE 数据库中目标栅格数据集；

②打开待入库的栅格数据文件(用到辅助函数 OpenFileRaster())，得到栅格数据集接口对象 IRasterDataset。然后，用此接口对象的 CreateDefaultRaster()函数创建 IRaster 接口对象；

③创建栅格数据集装载接口对象 IRasterLoader；

④加载 Raster 到 GDB RasterDataset 中(使用 IRasterLoader 的 Load 函数)。

源代码如下：

```
    public bool LoadRasterIntoDbDataset ( string fileFullName, string
strDbRasterDatasetName)
    {
```

```
        if (m_pWorkspace==null)
            return false;

        //打开 SDERasterDataset
        IRasterWorkspaceEx ipSdeRasterWs=m_pWorkspace as IRasterWork-
spaceEx;
        IRasterDataset ipRasterDataset =
                                          ipSdeRasterWs.OpenRasterDataset
(strDbRasterDatasetName);

        //打开文件中的 Raster
        IRasterDataset ipFileRasterDataset=this.OpenRasterFromFile(file-
FullName);
        if (ipFileRasterDataset==null) return false;

        IRaster ipRaster=ipFileRasterDataset.CreateDefaultRaster();

        //创建栅格数据集装载接口对象 IRasterLoader
        IRasterLoader ipRasterLoader=new RasterLoader();
        ipRasterLoader.Background=0;
        ipRasterLoader.PixelAlignmentTolerance=0.5;
        ipRasterLoader.MosaicColormapMode=rstMosaicColormapMode.MM_LAST;

        //加载 Raster 到 RasterDataset 中
        ipRasterLoader.Load(ipRasterDataset,ipRaster);
        return true;
    }
```

3. 辅助函数：OpenRasterFromFile()

代码如下：

```
    private IRasterDataset OpenRasterFromFile(string fileName)
    {
        if (fileName.Length==0) return null;
        int lastIndex=fileName.LastIndexOf(@"\");
        string strDir=fileName.Substring(0,lastIndex);
        string strFile=fileName.Substring(lastIndex+1);

        IWorkspaceFactory ipFileWsFactory=new RasterWorkspaceFactory-
Class();
```

```
    IWorkspace ipFileWSpace = ipFileWsFactory.OpenFromFile(strDir,0);;

    IRasterWorkspace ipFileRasterWs = ipFileWSpace as IRasterWorkspace;
    IRasterDataset ipRasterDataset = null;
    ipRasterDataset = ipFileRasterWs.OpenRasterDataset(strFile);

    return ipRasterDataset;
}
```

20.4　功能调用

1. 添加菜单项

在 DbManagerDockFrm 的浮动菜单(contextMenuDbManager)上，添加 New、Load 菜单项，以及相应的子菜单，见表 20-2。

表 20-2　菜单项及相应的子菜单

菜单项	子菜单	备　注
New	Feature Dataset	创建矢量数据集(数据库节点有效)
	Feature Class	创建要素类(矢量数据集节点有效)
	Raster Dataset	创建栅格数据集(数据库节点有效)
Load	Feature Class	装载要素类(在已有要素类结构上)
	Raster Dataset	装载栅格数据(在已有栅格数据集结构上)

2. 子菜单响应函数

相应的子菜单响应函数分别调用到 CreateDatasetFrm(可创建矢量和栅格两种数据集)，CreateFeatureClassFrm，LoadOpsClass(装载操作)等功能类。实现代码如下：

```
//创建矢量数据集
private void newfeatureDatasetToolStripMenuItem_Click(object sender,EventArgs e)
{
    esriDatasetType dsType = (esriDatasetType)m_pTreeNode.Tag;
    //限于数据库节点
    if (dsType = = esriDatasetType.esriDTSchematicDataset)
    {
        CreateDatasetFrm frm = new CreateDatasetFrm(m_pWorkspace,esriDatasetType.esriDTFeatureDataset);
        if (frm.ShowDialog() = = DialogResult.OK)
```

```
            {
                string NameOfNode=m_pTreeNode.FullPath+".DBO."+frm._
NameOfDataset;
                TreeNode subNode=m_pTreeNode.Nodes.Add(NameOfNode);
                subNode.Tag=esriDatasetType.esriDTFeatureDataset;
                subNode.ImageIndex=2;
                subNode.SelectedImageIndex=2;
            }
        }
    }
    //创建要素类
    private void newfeatureClassToolStripMenuItem_Click(object sender,
EventArgs e)
    {
        esriDatasetType dsType=(esriDatasetType)m_pTreeNode.Tag;
        //限于要素数据集节点
        if (dsType==esriDatasetType.esriDTFeatureDataset)
        {
            string dsName=m_pTreeNode.FullPath;
            int lastIndex=dsName.LastIndexOf(@"\");
            dsName=dsName.Substring(lastIndex+1);
            CreateFeatureClassFrm frm=new CreateFeatureClassFrm(m_pWork-
space,dsName);
            if(frm.ShowDialog()==DialogResult.OK)
            {
                string NameOfNode=m_pTreeNode.Parent.FullPath+".DBO."+
frm._NameOf-FeatureClass;
                TreeNode subNode=m_pTreeNode.Nodes.Add(NameOfNode);
                subNode.Tag=esriDatasetType.esriDTFeatureClass;
                subNode.ImageIndex=6;
                subNode.SelectedImageIndex=6;
            }
        }
    }
    //创建栅格数据集
    private void newRasterDatasetToolStripMenuItem_Click(object sender,
EventArgs e)
    {
```

```
        esriDatasetType dsType=(esriDatasetType)m_pTreeNode.Tag;
        //限于数据库节点
        if (dsType==esriDatasetType.esriDTSchematicDataset)
        {
            CreateDatasetFrm frm=new CreateDatasetFrm(m_pWorkspace,esri-
DatasetType.esriDTRasterDataset);
            if (frm.ShowDialog()==DialogResult.OK)
            {
                string NameOfNode=m_pTreeNode.FullPath +".DBO." + frm._
NameOfDataset;
                TreeNode subNode=m_pTreeNode.Nodes.Add(NameOfNode);
                subNode.Tag=esriDatasetType.esriDTRasterDataset;
                subNode.ImageIndex=3;
                subNode.SelectedImageIndex=3;
            }
        }
    }
    //装载要素类
    private void loadfeatureClassToolStripMenuItem1_Click(object send-
er,EventArgs e)
    {
        esriDatasetType dsType=(esriDatasetType)m_pTreeNode.Tag;
        //限于要素数据集节点
        if (dsType==esriDatasetType.esriDTFeatureClass)
        {
            OpenFileDialog dlg=new OpenFileDialog();
            dlg.Filter="shp file (*.shp) |*.shp";
            dlg.Title="打开矢量数据";
            dlg.Multiselect=false;
            if (dlg.ShowDialog()==DialogResult.OK)
            {
                string fileName=dlg.FileName;
                int lastIndex=fileName.LastIndexOf(@ "\");
                string shpFilePath=fileName.Substring(0,lastIndex);
                string shpFileName=fileName.Substring(lastIndex+1);

                string dsName=m_pTreeNode.FullPath;
                lastIndex=dsName.LastIndexOf(@ "\");
```

```
                dsName=dsName.Substring(lastIndex+1);

                LoadOpsClass loader=new LoadOpsClass(m_pWorkspace);
                loader.LoadShapefileToDatabase(shpFilePath,shpFileName,
dsName);
            }
        }
    }
    //装载栅格数据
    private void loadRasterToolStripMenuItem_Click(object sender,Even-
tArgs e)
    {
        esriDatasetType dsType=(esriDatasetType)m_pTreeNode.Tag;
        //限于栅格数据集节点
        if (dsType==esriDatasetType.esriDTRasterDataset)
        {
            OpenFileDialog dlg=new OpenFileDialog();
            dlg.Filter="tif file (*.tif) |*.* ";
            dlg.Title="打开矢量数据";
            dlg.Multiselect=false;
            if (dlg.ShowDialog()==DialogResult.OK)
            {
                string rasterfileName=dlg.FileName;

                string dsName=m_pTreeNode.FullPath;
                int lastIndex=dsName.LastIndexOf(@ "\");
                dsName=dsName.Substring(lastIndex+1);

                LoadOpsClass loader=new LoadOpsClass(m_pWorkspace);
                loader.LoadRasterIntoDbDataset(rasterfileName,dsName);
            }
        }
    }
```

20.5 运行测试

功能实现后，利用第 19 章的数据库连接功能，先连接 SQLExpress 数据库，然后对数据使用 New 菜单的操作创建相关存储结构，最后用 Load 操作将数据入库。

第 21 章　三 维 展 示

21.1　知识要点

ArcGIS Engine 可用于三维场景展示的控件是 GlobeControl 和 SceneControl(对应 ArcGIS Desktop 中 ArcGlobe 和 ArcScene)，二者在三维场景展示中适用的情况有所不同：前者适合大范围的数据展示，后者适合于小范围内精细场景刻画。但在编程技术上基本类同，本章采用 SceneControl 编程进行介绍。

SceneControl 编程常用的接口有 IScene、ISceneGraph、ISceneViewer、I3DViewer、ICamera 等接口，除此之外，IGraphicsLayers3D、I3DProperties 也经常使用。为帮助理解这些接口，我们列出其与 MapControl 编程的对应关系：

①SceneControl < = = > MapControl；SxDocment < = = >MxDocment；Scene < = = > Map：Scene 之于 SceneControl 如同 Map 之于 MapControl。在一个 SceneControl 中，只有一个 Scene 对象，Scene 是许多图层的集合。

②SceneGraph　< = = > Dispay：SceneGraph 可以看作是一个三维世界，它负责处理大部分的三维渲染操作，并且使绘图更有效率。

③ISceneViewer、I3DViewer < = = > IActiveView：ISceneViewer 实现和 Viewer 相关的功能(影像导出、快照)。因此，可以把 I3DViewer 看作是功能增强版的 ISceneViewer。

④IGraphicsContainer3D < = = >IGraphicsContainer。

⑤ICamera 定义了每一个 3D Viewer 的视角，方向和位置等，相当于提供摄像机视角的功能。

21.2　功能描述

实现 ArcScene 类似功能界面：①加载数据；②3D 图层属性设置；③3D 场景浏览。

21.3　功能实现

21.3.1　建立 3D 应用程序框架

1. 使用 GlobeControl Application 向导新建 ArcGIS Engine 应用程序

①将 GlobeControl 控件更换为 SceneControl 控件。

②将 TOCControl 和 ToobarControl 的 Buddy 属性设置为 SceneControl。

③修改相应代码：将 MainForm 文件中 Globe/globe 替换为 Scene/scene。

④删除 m_ sceneViewUtil 变量及相关代码。

2. 添加浏览工具

第一步：清空 ToolbarControl 控件的所有按钮。

第二步：进入"ToolbarControl"属性对话框中的"items"页面，并单击【Add…】按钮。弹出 Control Commands 对话框，在 Control Commands 对话框中选中"Category"列表框中的"Scene"选项，在"Commands"列表中就会出现与"Scene"关联的命令，如图 21-1 所示，双击命令就可以将该命令加入到"ToolbarControl"工具条中，如图 21-2 所示。

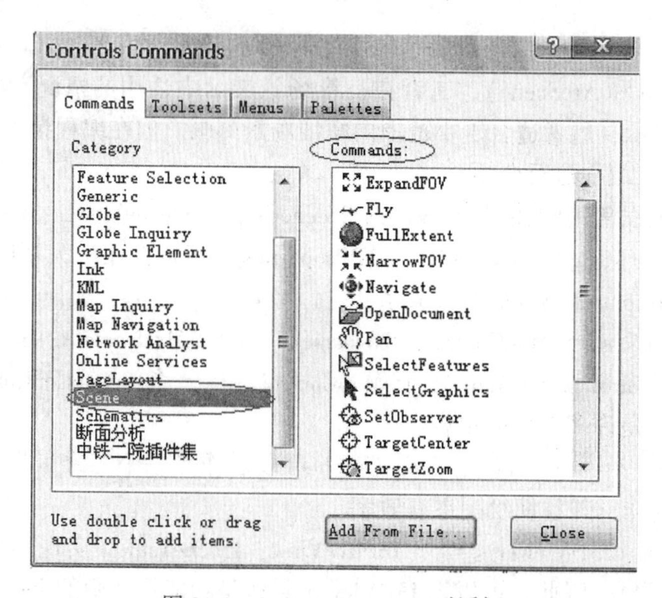

图 21-1 Control Commands 对话框

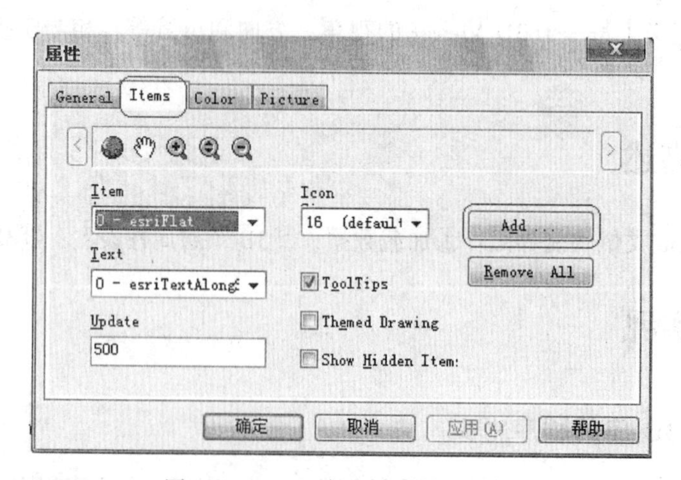

图 21-2 ToolbarControl 属性对话框

3. 修改 OnMouseMove 事件响应函数

代码如下：

```
private void axSceneControl1_OnMouseMove(object sender,ISceneCon-
trolEvents_OnMouseMoveEvent e)
{
    ISceneGraph pSceneGraph=m_sceneControl.SceneGraph;
    int px=e.x;
    int py=e.y;
    IPoint point=null;
    object pOwner;
    object pObject;
    pSceneGraph.Locate(this.axSceneControl1.SceneViewer,px,py,
            esriScenePickMode.esriScenePickAll,true,out point,out
pOwner,out pObject);
    if (point!=null)
    {
        MessageBox.Show(point.X+"_"+point.Y+"_"+point.Z);
    }
}
```

21.3.2　添加数据加载函数

1. 加载栅格数据

代码如下：

```
//向工程中添加栅格数据
private void addRasterLayerToolStripMenuItem_Click(object sender,
EventArgs e)
{
    //取消文件过滤
    mOpenFileDialog.Filter="所有文件|*.*";
    //打开文件对话框打开事件
    if (mOpenFileDialog.ShowDialog()==DialogResult.OK)
    {
        //从打开对话框中得到打开文件的全路径
        string sFileName=mOpenFileDialog.FileName;
        //新建栅格图层
        IRasterLayer pRasterLayer=new RasterLayerClass();
        //创建栅格图层
        pRasterLayer.CreateFromFilePath(sFileName);
```

```
            //将图层加入到控件中
            m_sceneControl.Scene.AddLayer(pRasterLayer,true);
            //设置摄像机视角范围
            SetCameraExtent(pRasterLayer.VisibleExtent);
            //刷新
            mSceneControl.Refresh();
        }
    }
```

2. 加载 Tin 数据

代码如下:

```
    private void addTinLayerToolStripMenuItem_Click(object sender,
EventArgs e)
    {
        IScene pScene=m_sceneControl.Scene;
        ISceneGraph pSceneGraph=pScene.SceneGraph;
        if (this.folderBrowserDialog1.ShowDialog()= =DialogResult.OK)
        {
            string tinPath=this.folderBrowserDialog1.SelectedPath;
            FileInfo fileInfo=new FileInfo(tinPath);
            IWorkspaceFactory tinWorkspaceFactory=new TinWorkspaceFac-
toryClass();
            if (tinWorkspaceFactory.IsWorkspace(fileInfo.DirectoryName))
            {
                //打开 Tin 工作空间
                ITinWorkspace tinWorkspace;
                tinWorkspace = tinWorkspaceFactory.OpenFromFile(fileInfo.
DirectoryName,0) as ITinWorkspace;
                //打开 Tin 数据集
                ITin tin=tinWorkspace.OpenTin(fileInfo.Name);
                //创建 Tin 图层
                ITinLayer tinLayer=new TinLayerClass();
                tinLayer.Dataset=tin;
                tinLayer.Visible=false;
                //将图层加入到控件中
                pScene.AddLayer(tinLayer as ILayer,true);
                //设置摄像机视角范围
                SetCameraExtent(tinLayer.AreaOfInterest);
                //刷新
```

```
                pSceneGraph.RefreshViewers();
            }
        }
    }
```

3. 加载矢量数据

代码如下：

```
    private void addFeatureLayerToolStripMenuItem _ Click ( object
sender,EventArgs e)
    {
        IScene pScene=m_sceneControl.Scene;
        ISceneGraph pSceneGraph=pScene.SceneGraph;

        this.openFileDialog1.Title="Feature Layer";
        this.openFileDialog1.DefaultExt=".shp";
        this.openFileDialog1.Filter="(*.shp) |*.shp";
        if (this.openFileDialog1.ShowDialog()= =DialogResult.OK)
        {
            string strPathName=this.openFileDialog1.FileName;
            string strPath=strPathName.Substring(0,strPathName.LastIndexOf
('\\'));
            string strFileName=strPathName.Substring(strPath.Length+
1,strPathName.Length-(strPath.Length+1));

            IWorkspaceFactory fwsf=new ShapefileWorkspaceFactoryClass
();
            IFeatureWorkspace featureWorkspace;
            if (fwsf.IsWorkspace(strPath))
            {
                //打开 shp 工作空间
                featureWorkspace=fwsf.OpenFromFile(strPath,0) as IFea-
tureWorkspace;
                //打开 shp 数据集
                IFeatureClass pFeatureClass=featureWorkspace.OpenFea-
tureClass(strFileName);
                //创建 shp 图层
                IFeatureLayer featureLayer=new FeatureLayerClass();
                featureLayer.FeatureClass=pFeatureClass;
                featureLayer.Name=strFileName;
```

```
                //配置渲染器
                IGeoFeatureLayer pGeoLayer = featureLayer as IGeoFeature-
Layer;
                pGeoLayer.Renderer = CreateFeatureRenderer(0,255,0);

                //将图层加入到控件中
                pScene.AddLayer( featureLayer as ILayer,true);
                //pScene.ExaggerationFactor = 6;
                //刷新
                pSceneGraph.RefreshViewers();
            }
        }
    }
```

21.3.3 建立属性设置窗体

1. 新建显示属性设置对话框类

设计界面如图 21-3 所示。

图 21-3 设计界面

对话框取名为：PropertySetingFrm，包括如下控件，详见表 21-1。

表 21-1 控件类型及名称

控件类型	控件名称	备　注
ComboBox	cbxLayer	基面
TextBox	txtTransParency	透明度

续表

控件类型	控件名称	备　　注
TextBox	txtZFactor	Z 轴缩放因子
TextBox	txtLayerOffset	Z 偏移
Button	btnOK	确定
Button	btnCancel	取消

(1)构造函数为两个私有成员赋值

代码如下:

```
private IScene m_pScene=null;
private ILayer m_pLayer=null;
public PropertySettingFrm(IScene pScene,ILayer layer)
{
    m_pScene=pScene;
    m_pLayer=layer;
    InitializeComponent();
}
```

(2)加载事件响应函数填充 cbxLayer

代码如下:

```
private void PropertySettingFrm_Load(object sender,EventArgs e)
{
    //load all the feature layers in the map to the layers combo
    IEnumLayer layers=GetLayers();
    layers.Reset();
    ILayer layer=null;
    while ((layer=layers.Next())!=null)
    {
        cboLayers.Items.Add(layer.Name);
    }
    //select the first layer
    if (cboLayers.Items.Count > 0)
        cboLayers.SelectedIndex=0;
}
```

(3)OK 事件响应函数执行配置

代码如下:

```
private void btnOK_Click(object sender,EventArgs e)
{
```

```
        double offset=double.Parse(this.txtOffset.Text.ToString());
        short percentage=short.Parse(this.txtTransparency.Text.ToString());
        short ZFactor=short.Parse(this.txtZFactor.Text.ToString());

        //取得基准面
        string baseLayerName=this.cboLayers.SelectedItem.ToString();
        ILayer baseLayer=GetLayer(baseLayerName);

        //设置高程基准面
        set3DLayerBaseHight(baseLayer,m_pLayer);
        //设置Z缩放因子
        set3DLayerZFactor(m_pLayer,ZFactor);
        //设置显示透明度
        setTransparency(m_pLayer,percentage);
        //设置图层Z轴偏移
        set3DLayerOffset(m_pLayer,offset);

        this.Close();
    }
```

2. 功能实现

(1)set3DLayerBaseHight(…) 函数

本函数设置高程基准面，步骤如下：

①根据指定基准图层，确定基准表面；

②从待设图层中确定 3D 属性接口对象；

③为接口赋参数：BaseOption 指定 esriBaseSurface，BaseSurface 指定基准面；

④应用 3D 属性到待设图层。

具体代码如下：

```
public void set3DLayerBaseHight(ILayer baseLayer,ILayer setingLayer)
{
    //获取基准面:
    ISurface surface=getSurface(baseLayer);

    //获取待设置图层的 3D 属性接口
    I3DProperties properties=get3DPropertiesFromLayer(setingLayer);

    //为接口赋参数
    properties.BaseOption=esriBaseOption.esriBaseSurface;
```

```
    properties.BaseSurface=surface;

    //应用到图层
    properties.Apply3DProperties(setingLayer);
}
```

(2) set3DLayerZFactor(...) 函数

本函数设置高程方向缩放因子，代码如下：

```
public void set3DLayerZFactor(ILayer pLayer,double ZFactor)
{
    I3DProperties p3DProps=get3DPropertiesFromLayer(pLayer);
    if (p3DProps!=null)
    {
        p3DProps.ZFactor=ZFactor;
        //设置高程缩放因子
        p3DProps.Apply3DProperties(pLayer);
    }
}
```

(3) setTransparency(...) 函数和 set3DLayerOffset(...) 函数

这两个函数分别设置显示透明度和图层 Z 轴偏移量，代码如下：

```
public void set3DLayerOffset(ILayer pLayer,double offset)
{
    I3DProperties p3DProps=get3DPropertiesFromLayer(pLayer);
    if (p3DProps!=null)
    {
        p3DProps.OffsetExpressionString=offset.ToString();
        p3DProps.Apply3DProperties(pLayer);
    }
}

public void setTransparency(ILayer setingLayer,short percentage)
{
    ILayerEffects lyEffects=setingLayer as ILayerEffects;
    lyEffects.Transparency=percentage;
}
```

(4) 辅助函数

代码如下：

```
//获取基准面
private ISurface getSurface(ILayer baseLayer)
```

```
{
    ISurface surface=null;
    if (baseLayer is IRasterLayer)
    {
        //获取栅格图层的 IRaster 接口
        IRasterLayer rLayer=baseLayer as IRasterLayer;
        IRaster raster=(IRaster)rLayer.Raster;

        //栅格数据的第一波段
        IRasterBandCollection rasterbands = raster as IRasterBandCol-
lection;
        IRasterBand rasterband=rasterbands.Item(0);

        //创建栅格表面
        IRasterSurface rsurface=new RasterSurface();
        rsurface.RasterBand=rasterband;
        surface=rsurface as ISurface;
    }
    else if (baseLayer is ITinLayer)
    {
        ITinLayer tinLayer=baseLayer as ITinLayer;
        ITinAdvanced tinAdvanced=tinLayer.Dataset as ITinAdvanced;
        surface=tinAdvanced.Surface;
    }

    return surface;
}

//获取待设置图层的 3D 属性接口
private I3DProperties get3DPropertiesFromLayer(ILayer pLayer)
{
    I3DProperties p3DProperties=null;
    ILayerExtensions pLayerExtensions=pLayer as ILayerExtensions;
    if (pLayerExtensions!=null)
    {
        for (int i=0; i < pLayerExtensions.ExtensionCount; i++)
        {
            p3DProperties = pLayerExtensions.get_Extension(i) as
```

```
I3DProperties;
                if (p3DProperties!=null)
                    break;
            }
        }

        if (p3DProperties==null)
        {
            p3DProperties=new Raster3DPropertiesClass();
            ILayerExtensions pLayerEx=(pLayer as ILayerExtensions);
            pLayerEx.AddExtension(p3DProperties);
        }

        return p3DProperties;
    }
    //根据层名获取图层
    private ILayer GetLayer(string layerName)
    {
        //get the layers from the maps
        IEnumLayer layers=GetLayers();
        layers.Reset();

        ILayer layer=null;
        while ((layer=layers.Next())!=null)
        {
            if (layer.Name==layerName)
                return layer;
        }

        return null;
    }
    //获取 Scene 中所有图层
    private IEnumLayer GetLayers()
    {
        UID uid=new UIDClass();
        //uid.Value="{40A9E885-5533-11d0-98BE-00805F7CED21}";
        uid.Value="{D02371C7-35F7-11D2-B1F2-00C04F8EDEFF}"; //Raster-
Layer
```

```
        IEnumLayer layers=m_pScene.get_Layers(uid,true);

        return layers;
    }
```

3. 功能调用

①在 TOCControl 中添加 MouseDown 响应函数；并添加 m_tocRightLayer，m_tocRightLegend 私有成员，记录右键点击的图层或图例。

源代码如下：

```
    private ILayer m_tocRightLayer=null;
    private ILegendClass m_tocRightLegend=null;
    private void axTOCControl1_OnMouseDown(object sender,
                                                ITOCControlEvents_OnMouse-
DownEvent e)
    {
        esriTOCControlItem itemType=esriTOCControlItem.esriTOCContro-
lItemNone;
        IBasicMap basicMap=null;
        ILayer layer=null;
        object unk=null;
        object data=null;
        this.axTOCControl1.HitTest(e.x,e.y,ref itemType,ref basicMap,
ref layer,ref unk,ref data);
        if (e.button==2)
        {
            switch (itemType)
            {
                case esriTOCControlItem.esriTOCControlItemLayer:
                    this.m_tocRightLayer=layer;
                    this.m_tocRightLegend=null;
                    this.contextMenuTOCLyr.Show(this.axTOCControl1,e.x,e.y);
                    break;
                case esriTOCControlItem.esriTOCControlItemLegendClass:
                    this.m_tocRightLayer=layer;
                    this.m_tocRightLegend=((ILegendGroup)unk).get_Class((int)
data);
                    this.contextMenuTOCLyr.Show(this.axTOCControl1,e.x,e.y);
                    break;
                case esriTOCControlItem.esriTOCControlItemMap:
```

```
                    //this.contextMenuTOCMap.Show(this.axTOCControl1,
e.x,e.y);
                    break;
            }
        }
    }
```

②添加浮动菜单 contextMenuTOCLyr，并为其添加菜单项 Property。在其响应函数调用 3D 属性设置对话框类，代码如下：

```
    private void propertysToolStripMenuItem_Click(object sender,Even-
tArgs e)
    {
        PropertySettingFrm frm=new PropertySettingFrm(m_sceneControl.
Scene,m_tocRightLayer);
        frm.ShowDialog();
    }
```

21.4　运行测试

按下 F5 键，编译运行程序。

第 22 章　创建 TIN

22.1　知识要点

TIN 是 GIS 典型的表面模型，通常由矢量数据创建。

ArcGIS Engine 创建 TIN 需要用到 ITinEdit 接口，通过接口的 AddFromFeatureClass 方法添加供创建 TIN 的矢量数据图源，最后都调用 ITinEdit 接口的 SaveAs( )方法将创建的 TIN 数据保存。

TIN 向 DEM 转换，可采用 GP 工具 TinRaster 实现。

22.2　功能描述

单击【Spatial Analysis】页【Create TIN】按钮，弹出如下 Create TIN 对话框：

选取“输入图层”，“高程字段”，以及参与类型(质点/软线/硬线/软裁剪/硬裁剪/软替换/硬替换)，点击【+】按钮将其添加到列表框。添加完毕，点击【生成 TIN】按钮即将生成 TIN 数据文件保存在输出图层文本框指定的位置。之后，可点击【转换 DEM】按钮，将 TIN 转换为 DEM 数据。操作界面如图 22-1 所示。

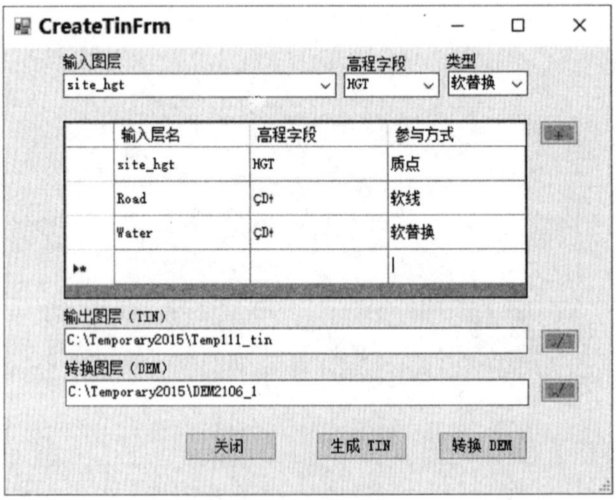

图 22-1　操作界面

22.3　功能实现

22.3.1　新建功能窗体

1. 界面设计

项目中添加一个新的窗体，名称为“CreateTINFrm”，Name 属性设置为“Create TIN”，添加 3 个 ComboBox、2 个 TextBox，1 个 DataGridView、6 个 Button 控件。

控件属性设置详见表 22-1。

表 22-1　控件属性设置

控件类型	Name 属性	控件说明	备　注
ComBox	cbxInLayer	输入图层	
ComBox	cbxFields	高程字段	
ComBox	cbxTINType	参与类型	Items 集合中填充：质点/软线/硬线/软裁剪/硬裁剪/软替换/硬替换
DataGridView	dataGridView1	参与计算图层列表	
TextBox	txtOutTinLayer	输出图层(TIN)	
TextBox	txtOutDemLayer	转换图层(DEM)	
Button	btnTinGenerate	生成 TIN	
Button	btnDemConverter	转换 DEM	
Button	btnTinBrowser	TIN 文件名	
Button	btnDemBrowser	DEM 文件名	
Button	btnAdd	添加要素到表格	
Button	btnCancel	取消	

2. 类结构设计

添加如下引用代码，修改类定义代码：

```
public partial class CreateTinFrm:Form
{
    ISceneControl m_sceneControl = null;
    IEnvelope m_pEnvelope = null;
    ISpatialReference m_pSpatialReference = null;
    public CreateTinFrm( ISceneControl sceneControl )
```

```
    {
        InitializeComponent();
        m_sceneControl=sceneControl;
    }
    //Load 事件响应函数
    private void CreateTinFrm_Load(object sender,EventArgs e)
    //输入图层名变化响应函数
    private void cbxInLayer_SelectedIndexChanged(object sender,
EventArgs e)
    //输入图层添加按钮响应函数
    private void btnAdd_Click(object sender,EventArgs e)

    //TIN 文件名按钮响应函数
    private void btnTinBrowser_Click(object sender,EventArgs e)
    //DEM 文件按钮响应函数
    private void btnDemBrowser_Click(object sender,EventArgs e)
    //转换 DEM 按钮响应函数
    private void btnDemConverter_Click(object sender,EventArgs e)
    //创建 Tin 按钮响应函数
    private void btnTinGenerate_Click(object sender,EventArgs e)

    //取消按钮响应函数
    private void btnCancel_Click(object sender,EventArgs e)

    //若干辅助函数
    ……
}
```

22.3.2 消息响应函数

1. 载入响应函数 CreateTinFrm_Load()

①创建 DataGridView 空表，共三列：输入层名、高程字段、参与方式。

②设置输出文件的默认输出路径，这里我们将默认输出路径设为系统临时目录。

③如果 m_sceneControl 中图层数大于 0，用它们填充 cbxInLayer 下拉框，然后将下拉框索引设置为第一项。

```
private void CreateTinFrm_Load(object sender,EventArgs e)
{
    //创建 DataGrid 表头
    DataTable pDataTable=new DataTable();
```

```
        DataColumn lyerValue = new DataColumn ( " 输入层名", System. Type.
GetType(" System.String"));
        DataColumn fldValue=new DataColumn("高程字段",System.Type.Get-
Type(" System.String"));
        DataColumn typeValue = new DataColumn ( " 参与方式", System. Type.
GetType(" System.String"));
        pDataTable.Columns.Add(lyerValue);
        pDataTable.Columns.Add(fldValue);
        pDataTable.Columns.Add(typeValue);
        this.dataGridView1.DataSource=pDataTable;

        txtOutTinLayer.Text=@"C:\Temporary2015\Temp111_tin";
        txtOutDemLayer.Text=@"C:\Temporary2015\DEM2106_1.GRID";

        //得到当前场景中所有图层
        int nCount=m_sceneControl.Scene.LayerCount;
        if (nCount <=0)//没有图层的情况
        {
            MessageBox.Show("场景中没有图层,请加入图层");
            return;
        }

        //将所有图层名添加到输入图层下拉框
        cbxInLayer.Items.Clear();
        ILayer pLayer=null;
        for (int i=0; i < nCount; i++)
        {
            pLayer=m_sceneControl.Scene.get_Layer(i);
            if(pLayer is IFeatureLayer)
                cbxInLayer.Items.Add(pLayer.Name);
        }
        //将下拉框索引设置为第一项
        cbxInLayer.SelectedIndex=0;
    }
```

2. 输入图层变化响应函数 cbxInLayer_SelectedIndexChanged()

图层名称发生变化时，将该图层的字段加入到 cbxFields 下拉框中。代码如下：

```
    private void cbxInLayer_SelectedIndexChanged(object sender,Even-
tArgs e)
```

```
{
    cbxFields.Items.Clear();
    IFeatureLayer pFeatureLayer=getLayerFromName(cbxInLayer.Text);
    if (pFeatureLayer!=null)//判断是否找到
    {
        IFields pFields=pFeatureLayer.FeatureClass.Fields;
        int nCount=pFields.FieldCount;
        //将该图层中可用的字段添加到 Feild Combox 中
        for (int i=0; i < nCount; i++)
        {
            IField pField=pFields.get_Field(i);
            if (pField.Type==esriFieldType.esriFieldTypeInteger ||
                pField.Type==esriFieldType.esriFieldTypeDouble ||
                pField.Type==esriFieldType.esriFieldTypeSingle)//
判断可用类型
            {

                cbxFields.Items.Add(pField.Name);
            }
        }
    }
    cbxFields.SelectedIndex=0;
}
```

3. 输入图层添加按钮响应函数 btnAdd_Click()

根据输入图层/高程字段/参与类型三个下拉框选定内容为 DataGridView 数据源 Datatable 添加一行，同时修订范围和参考系成员变量的值。代码如下：

```
private void btnAdd_Click(object sender,EventArgs e)
{
    //为 DataGridView 数据源 Datatable 添加一行
    DataTable pDataTable=this.dataGridView1.DataSource as DataTable;
    DataRow pRow=pDataTable.NewRow();

    pRow[0]=cbxInLayer.SelectedItem.ToString();
    pRow[1]=cbxFields.SelectedItem.ToString();
    pRow[2]=cbxTINType.SelectedItem.ToString();
    pDataTable.Rows.Add(pRow);

    this.dataGridView1.DataSource=pDataTable;
```

```
        this.dataGridView1.Refresh();

        //修订范围和参考系
        IFeatureLayer pFeatureLayer = getLayerFromName(cbxInLayer.Se-
lectedItem.ToString());
        if (pFeatureLayer!=null)
        {
            IGeoDataset pGeoDataset=pFeatureLayer as IGeoDataset;
            IEnvelope pEnvelope=pGeoDataset.Extent;
            if (m_pEnvelope==null)
                m_pEnvelope=pEnvelope;
            else
                m_pEnvelope.Union( pEnvelope );

            if(m_pSpatialReference==null)
                m_pSpatialReference=pGeoDataset.SpatialReference;
        }
    }
```

4. 输出路径设置响应函数 btnTinBrowser_Click()、btnDemBrowser_Click()

TIN 输出路径设置由 FolderBrowserDialog 实现，DEM 输出文件设置由 SaveFileDialog 实现，添加代码如下：

```
    private void btnTinBrowser_Click(object sender,EventArgs e)
    {
        //set the output layer
        FolderBrowserDialog folderDlg=new FolderBrowserDialog();
        DialogResult dr=folderDlg.ShowDialog();
        if (dr==DialogResult.OK)
            txtOutTinLayer.Text=folderDlg.SelectedPath;
    }

    private void btnDemBrowser_Click(object sender,EventArgs e)
    {
        //set the output layer
        SaveFileDialog saveDlg=new SaveFileDialog();
        saveDlg.CheckPathExists=true;
        saveDlg.Filter="Tinfile (*.GRID) |*.GRID";
        saveDlg.OverwritePrompt=true;
        saveDlg.Title="Output Layer";
```

```
        saveDlg.RestoreDirectory = true;

        DialogResult dr = saveDlg.ShowDialog();
        if (dr == DialogResult.OK)
            txtOutDemLayer.Text = saveDlg.FileName;

    }
```

5. 生成 TIN 响应函数 btnTinGenerate_Click()

步骤如下：

①初始化 ITinEdit 接口。

②将 DataGridView 所列表中参与 TIN 计算信息添加到 ITinEdit 接口对象，此处用到 AddParticipateInfoToTin()函数。

③ITinAdvanced 接口 Save 函数将 TIN 结果保存到指定目录。

④创建 Tin 图层并将 Tin 图层加入到三维场景中。

代码如下：

```
    private void btnTinGenerate_Click(object sender,EventArgs e)
    {
        //初始化 Tin
        ITinEdit pTin = new TinClass();
        pTin.InitNew(m_pEnvelope);
        pTin.SetSpatialReference(m_pSpatialReference);

        //添加参与 Tin 计算信息
        DataTable pTable = (dataGridView1.DataSource) as DataTable;
        AddParticipateInfoToTin(pTin,pTable);

        //结果保存
        ITinAdvanced tinAdv = pTin as ITinAdvanced;
        object obj = Type.Missing;
        tinAdv.SaveAs(this.txtOutTinLayer.Text,obj);

        //创建 Tin 图层并将 Tin 图层加入到场景中去
        ITinLayer pTinLayer = new TinLayerClass();
        pTinLayer.Dataset = pTin as ITin;
        pTinLayer.Name = System.IO.Path.GetFileName(this.txtOutTinLay-
er.Text);
        m_sceneControl.Scene.AddLayer(pTinLayer,true);
    }
```

6. 转换 DEM 响应函数 btnDemConverter_ Click( )

代码如下:

```
private void btnDemConverter_Click(object sender,EventArgs e)
{
    try
    {
         TinToGrid(this.txtOutTinLayer.Text,this.txtOutDemLayer.
Text,"OBSERVATIONS 1500");
    }
    catch (Exception ex)
    {
        MessageBox.Show( ex.Message.ToString() );
    }
}
```

22.3.3 核心函数

1. AddParticipateInfoToTin( )函数

用 ITinEdit 的 AddFromFeatureClass( )函数，将 DataTable 中“参与图层”，“高层字段”，“参与类型”等信息添加到 ITinEdit 接口中。

代码如下:

```
private void AddParticipateInfoToTin(ITinEdit pTin,DataTable pT-
able)
{
    for (int i=0; i < pTable.Rows.Count; i++)
    {
        DataRow row=pTable.Rows[i];
        String lyrName=row[0].ToString();
        String fldName=row[1].ToString();
        String typeName=row[2].ToString();

        //寻找 Featurelayer
        IFeatureLayer pFeatureLayer=getLayerFromName(lyrName);
        if (pFeatureLayer==null)
            continue;

        //寻找高程字段
        IFeatureClass pFls=pFeatureLayer.FeatureClass;
        IFields pFields=pFls.Fields;
```

```
            int FieldIndex=pFields.FindField(fldName);
            IField pField=(FieldIndex > 0) ? pFields.get_Field(FieldIn-
dex):null;

            //准备参数
            esriTinSurfaceType pSurfaceType=getTinSurfaceType(typeName);
            object missing=Type.Missing;
            IQueryFilter pQueryFilter=null;

            //添加参与 TIN 的数据
            pTin.AddFromFeatureClass(pFls,pQueryFilter,pField,pField,
pSurfaceType,ref missing);
        }
    }
```

2. TinToGrid()函数

实现 TIN 向 DEM 转换的功能，本函数采用 GP 功能实现，代码如下：

```
    private void TinToGrid ( string tempBathyTIN, string strGridPath,
string samplingMethod)
    {
        string tinFolder=System.IO.Path.GetDirectoryName(tempBathyTIN);
        string tinName=System.IO.Path.GetFileName(tempBathyTIN);
        IWorkspaceFactory TinWF=new TinWorkspaceFactory();
        ITinWorkspace TinWK = TinWF.OpenFromFile(tinFolder,0) as ITin-
Workspace;
        ITinAdvanced2 tinAdv=TinWK.OpenTin(tinName) as ITinAdvanced2;

        ITinLayer pTinLayer=new TinLayerClass();
        pTinLayer.Dataset=tinAdv as ITin;
        try
        {
            Geoprocessor gp=new Geoprocessor();
            gp.OverwriteOutput=true;
            TinRaster tinToRaster=new TinRaster();
            tinToRaster.in_tin=pTinLayer;
            tinToRaster.out_raster=strGridPath;
            tinToRaster.data_type="FLOAT";
            tinToRaster.method="LINEAR";
            tinToRaster.sample_distance = samplingMethod; //"OBSERVATIONS
```

```
1500";
            gp.Execute(tinToRaster,null);
            MessageBox.Show("转换完成");
        }
        catch (Exception e)
        {
            MessageBox.Show("转换失败,原因如下:"+e.Message.ToString());
        }
    }
```

22.3.4 辅助函数

代码如下:

```
//将字符串转换为 TIN 计算的参与类型
private esriTinSurfaceType getTinSurfaceType(string typeName)
{
    esriTinSurfaceType pSurfaceType;
    switch (typeName)
    {
        case "质点":
            pSurfaceType=esriTinSurfaceType.esriTinMassPoint;
            break;
        case "软线":
            pSurfaceType=esriTinSurfaceType.esriTinSoftLine;
            break;
        case "硬线":
            pSurfaceType=esriTinSurfaceType.esriTinHardLine;
            break;
        case "软裁剪":
            pSurfaceType=esriTinSurfaceType.esriTinSoftClip;
            break;
        case "硬裁剪":
            pSurfaceType=esriTinSurfaceType.esriTinHardClip;
            break;
        case "软替换":
            pSurfaceType=esriTinSurfaceType.esriTinSoftReplace;
            break;
        case "硬替换":
            pSurfaceType=esriTinSurfaceType.esriTinHardReplace;
```

```
                break;
            default:
                pSurfaceType = esriTinSurfaceType.esriTinMassPoint;
                break;
        }

        return pSurfaceType;
    }

    private IFeatureLayer getLayerFromName(string layerName)
    {
        IFeatureLayer pLayer = null;
        //寻找名称为 layerName 的 FeatureLayer;
        for (int i = 0; i < m_sceneControl.Scene.LayerCount; i++)
        {
            pLayer = m_sceneControl.Scene.get_Layer(i) as IFeatureLayer;
            if (pLayer.Name == layerName)//找到了 layerName 的 Featurelayer
            {
                break;
            }
        }
        return pLayer;
    }

    //显示栅格结果
    private void ShowRasterResult(IGeoDataset geoDataset,string inter-
Type)
    {
        IRasterLayer rasterLayer = new RasterLayerClass();
        IRaster raster = new Raster();
        raster = (IRaster)geoDataset;
        rasterLayer.CreateFromRaster(raster);
        rasterLayer.Name = interType;

        m_sceneControl.Scene.AddLayer((ILayer)rasterLayer,true);
        m_sceneControl.Scene.SceneGraph.RefreshViewers();
    }
```

22.4　功能调用

在【Spatial Analysis】页上，添加【Create TIN】按钮。建立 Click 响应函数；

```
private void btnCreateTIN_Click(object sender,EventArgs e)
{
    CreateTinFrm frm=new CreateTinFrm(m_sceneControl);
    frm.ShowDialog();
}
```

22.5　编译测试

按下 F5 键，编译运行程序。

运行程序，点击按钮【Create TIN】，弹出分析窗口，添加分析图层，并设置输出文件路径和文件名。

第 23 章　表 面 分 析

23.1　知识要点

在 ArcGIS Engine 中，RasterSurfaceOpClass 类实现了栅格数据的表面分析。该类实现了两个主要的接口，分别是 IRasterAnalysisEnvironment 接口和 ISurfaceOp 接口。ISurfaceOp 接口包含栅格数据表面分析的所有方法，主要有：坡度分析(Slope)、坡向分析(Aspect)、生成等值线(Contour)、填挖方(CutFill)、山体阴影(HillShade)、曲率(Curvature)、可见性(Visibility)。

下面将介绍坡度、坡向、等值线、填挖方、山体阴影、曲率和可见性这几种常用的表面分析方法，其他方法请读者自行参阅 ArcGIS Engine 的帮助文档。

23.2　功能描述

单击【Spatial Analysis】页【DEM Analyst】按钮，弹出如图 23-1 所示表面分析对话框，即根据输入栅格图层，输入矢量图层，生成相应的 DEM 分析结果数据(输出栅格数据)。

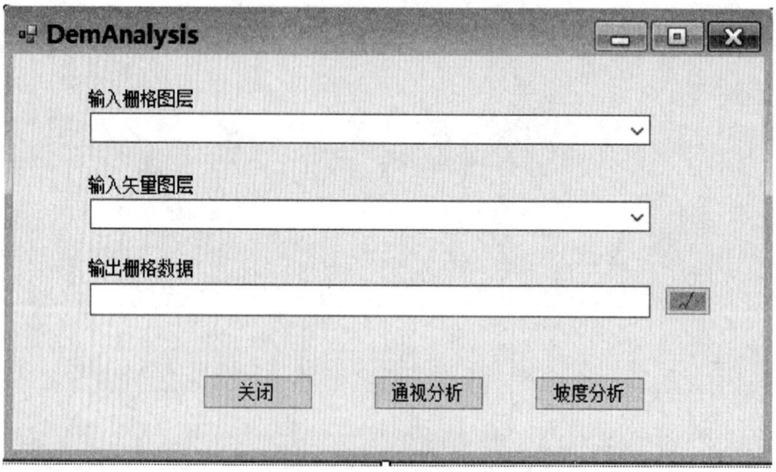

图 23-1　表面分析对话框

23.3　功能实现

23.3.1　新建功能窗体

1. 界面设计

项目中添加一个新的窗体，名称为“DemAnalysisFrm”，Name 属性设为“DemAnalysis”，添加 3 个 Label、2 个 ComboBox、1 个 TextBox、4 个 Button 控件。控件属性设置见表 23-1。

表 23-1　　控件属性设置

控件类型	Name 属性	控件说明	备　注
ComBox	cbxRasterLayer	输入栅格图层	
ComBox	cbxFeatureLayer	输入矢量图层	
TextBox	cbxOutRaster	输出栅格数据	
Button	btnSlopAnalyst	坡度分析按钮	
Button	btnVisibility	通视分析按钮	
Button	btnBrowser	文件浏览按钮	
Button	btnCancel	取消按钮	

2. 类结构设计

添加如下代码，修改类定义代码：

```
public partial class DemAnalysisFrm:Form
{
    ISceneControl m_sceneControl=null;
    //构造函数
    public DemAnalysisFrm(ISceneControl sceneControl)
    {
        InitializeComponent();
        m_sceneControl=sceneControl;
    }
    //装载响应函数
    private void DemAnalysisFrm_Load(object sender,EventArgs e)
    //坡度分析按钮相应函数
    private void btnSlopAnalyst_Click(object sender,EventArgs e)
    //可视分析按钮响应函数
    private void btnVisibility_Click(object sender,EventArgs e)
```

```
        //取消按钮响应函数
        private void btnCancel_Click(object sender,EventArgs e)
        //文件浏览响应函数
        private void btnBrowser_Click(object sender,EventArgs e)

        //核心功能函数============
        //坡度分析
        private void SlopeAnalyst(IRasterLayer rasterLayer,string out-
putFileName)
        //通视分析
        private void ViewAnalyst(IRasterLayer rasterLayer,IFeatureLayer
featureLayer,string outputFileName)
        //分析环境设置
        private void SetAnalysisEnvironment(IRasterAnalysisEnvironment
rasAnaEnv,IRaster costRaster)

        //辅助函数================
        //设置生成图层的工作空间
        private IWorkspace OpenRasterWorkspace(string outputFileName)
        //获取图层接口
        private ILayer getLayerFromName(string layerName)
        //显示栅格结果
        private void ShowRasterResult(IGeoDataset geoDataset,string in-
terType)
    }
```

23.3.2 消息响应函数

1. 载入响应函数 DemAnalysisFrm_Load()

用 ISceneControl.Scene 中的栅格图层名填充 cbxRasterLayer；矢量图层填充 cbxFeatureLayer。

```
    private void DemAnalysisFrm_Load(object sender,EventArgs e)
    {
        //得到当前场景中所有图层
        int nCount=m_sceneControl.Scene.LayerCount;
        if (nCount<=0)//没有图层的情况
        {
            MessageBox.Show("场景中没有图层,请加入图层");
            return;
```

```
    }

    ILayer pLayer=null;
    //将所有的图层的名称显示到复选框中
    cbxRasterLayer.Items.Clear();
    for (int i=0; i < nCount; i++)
    {
        pLayer=m_sceneControl.Scene.get_Layer(i);
        if(pLayer is IRasterLayer)
            cbxRasterLayer.Items.Add(pLayer.Name);
        else if(pLayer is IFeatureLayer)
            cbxFeatureLayer.Items.Add(pLayer.Name);
    }

    //将复选框设置为选中第一项
    cbxRasterLayer.SelectedIndex=0;
    cbxFeatureLayer.SelectedIndex=0;
}
```

2. 输出文件设置响应函数 btnOutRaster_Click()

输出文件设置由 SaveFileDialog 实现，添加代码如下：

```
//文件浏览响应函数
private void btnBrowser_Click(object sender,EventArgs e)
{
    //set the output layer
    SaveFileDialog saveDlg=new SaveFileDialog();
    saveDlg.CheckPathExists=true;
    saveDlg.Filter="Tinfile (*.GRID) |*.GRID";
    saveDlg.OverwritePrompt=true;
    saveDlg.Title="Output Layer";
    saveDlg.RestoreDirectory=true;

    DialogResult dr=saveDlg.ShowDialog();
    if (dr==DialogResult.OK)
        txtOutRaster.Text=saveDlg.FileName;

}
```

3. 分析响应函数 btnSlopAnalyst_Click()/ btnVisibility_Click()

分析响应函数：准备输入参数，调用相应的核心功能函数，代码如下：

```
//坡度分析按钮响应函数
private void btnSlopAnalyst_Click(object sender,EventArgs e)
{
    string layerName=cbxRasterLayer.SelectedItem.ToString();
    IRasterLayer rLayer=getLayerFromName( layerName) as IRasterLayer;

    this.SlopeAnalyst(rLayer,this.txtOutRaster.Text);
}
//可视分析按钮响应函数
private void btnVisibility_Click(object sender,EventArgs e)
{
    string rLayerName=cbxRasterLayer.SelectedItem.ToString();
    IRasterLayer rLayer = getLayerFromName(rLayerName) as IRaster-
Layer;
    string fLayerName=cbxFeatureLayer.SelectedItem.ToString();
    IFeatureLayer fLayer=getLayerFromName(fLayerName) as IFeature-
Layer;

    this.ViewAnalyst(rLayer,fLayer,this.txtOutRaster.Text);
}
```

23.3.3 核心函数

1. SlopeAnalyst()函数

SlopeAnalyst() 完成坡度分析工作，步骤如下：

①创建表面分析接口对象；

②设置分析环境；使用 SetAnalysisEnvironment()函数；

③坡度分析计算；调用 ISurfaceOp 接口的 Slope 方法；

④结果保存到指定位置。

具体代码如下：

```
//坡度分析
private void SlopeAnalyst(IRasterLayer rasterLayer,string output-
FileName)
{
    //IRasterLayer rasterLayer=layer as IRasterLayer;
    IGeoDataset rasDataset=rasterLayer.Raster as IGeoDataset;

    //创建表面分析接口对象
    ISurfaceOp surfaceOp=new RasterSurfaceOpClass();
```

```
        IRasterAnalysisEnvironment rasAnaEnv=surfaceOp as IRasterAnal-
ysisEnvironment;

        //设置分析环境
        SetAnalysisEnvironment(rasAnaEnv,rasterLayer.Raster);

        //坡度分析
        object zFactor=new object();
        IGeoDataset outputDataset=surfaceOp.Slope(rasDataset,
                                esriGeoAnalysisSlopeEnum. esriGeoAnalys-
isSlopeDegrees,ref zFactor);
        ShowRasterResult(outputDataset,"Slop");

        //保存结果
        string fileName=System.IO.Path.GetFileName(outputFileName);
        IRasterWorkspace workspace = OpenRasterWorkspace(outputFileName)
as IRasterWorkspace;
        IRasterBandCollection rasterBandCollection = outputDataset as
IRasterBandCollection;
        rasterBandCollection. SaveAs (fileName, workspace as IWorkspace,
"TIFF");
    }
```

2. ViewAnalyst()函数

ViewAnalyst()函数完成可视分析工作，步骤如下：

①创建表面分析接口对象；

②设置分析环境：使用 SetAnalysisEnvironment()函数；

③通视分析计算：调用 ISurfaceOp 接口的 Visibility 方法；

④结果保存到指定位置。

具体代码如下：

```
    //通视分析
    private void ViewAnalyst (IRasterLayer rasterLayer, IFeatureLayer
featureLayer,string outputFileName)
    {
        IGeoDataset featureDataset=featureLayer.FeatureClass as IGeo-
Dataset;
        IGeoDataset rasterGeoDataset=rasterLayer.Raster as IGeoDataset;
```

```
        //创建表面分析接口对象,设置分析环境
        ISurfaceOp surfaceOp=new RasterSurfaceOpClass();
        IRasterAnalysisEnvironment rasAnaEnv=surfaceOp as IRasterAnal-
ysisEnvironment;
        SetAnalysisEnvironment(rasAnaEnv,rasterLayer.Raster);

        //通视分析
        object zFactor=new object();
        IGeoDataset outputDataset=surfaceOp.Visibility(rasterGeoData-
set,featureDataset,esriGeoAnalysisVisibilityEnum.esriGeoAnalysisVis-
ibilityFrequency);
        ShowRasterResult(outputDataset,"Visibility");//显示

        //保存
        string fileName=System.IO.Path.GetFileName(outputFileName);
        IRasterWorkspace workspace = OpenRasterWorkspace(outputFileName)
as IRasterWorkspace;
        IRasterBandCollection rasterBandCollection=outputDataset as IRas-
terBandCollection;
        rasterBandCollection.SaveAs(fileName, workspace as IWorkspace,
"TIFF");
    }
```

3. SetAnalysisEnvironment()函数

本函数负责设置栅格分析环境，这是十分重要的一步，所有基于栅格分析的分析方法，都必须设置好分析环境，它是通过 IRasterAnalysisEnvironment 接口来实现的，所有进行栅格分析的组件都实现了 IRasterAnalysisEnvironment 接口。

代码如下：

```
    //分析环境设置
    private void SetAnalysisEnvironment(IRasterAnalysisEnvironment rasAnaEnv,
IRaster pRaster)
    {
        //设置生成图层的范围
        IGeoDataset rGeoDataset=pRaster as IGeoDataset;
        object extent=rGeoDataset.Extent;
        object missing=System.Reflection.Missing.Value;
        rasAnaEnv.SetExtent(esriRasterEnvSettingEnum.esriRasterEnvValue,
```

```
ref extent,ref missing);

        //设置生成图层的栅格大小
        IRasterProps rProps=pRaster as IRasterProps;
        IPnt p=rProps.MeanCellSize();
        object cellsize=(p.X+p.Y)/2;
        rasAnaEnv.SetCellSize(esriRasterEnvSettingEnum.esriRasterEn-
vMinOf,ref cellsize);
    }
```

23.3.4　辅助函数

代码如下：

```
    //生成图层的工作空间
    private IWorkspace OpenRasterWorkspace(string outputFileName)
    {
        IWorkspaceFactory wsf=new RasterWorkspaceFactoryClass();
        string outputPath = System.IO.Path.GetDirectoryName(output-
FileName);

        IWorkspace ws=wsf.OpenFromFile(outputPath,0);
        return ws;
    }

    //通过图层名得到图层
    private ILayer getLayerFromName(string layerName)
    {
        ILayer layer;
        IMap map=m_mapControl.Map;
        for (int i=0; i < map.LayerCount; i++)
        {
            layer=map.get_Layer(i);
            if (layerName==layer.Name)
                return layer;
        }
        return null;
    }

    //显示栅格结果
```

```
    private void ShowRasterResult(IGeoDataset geoDataset,string interType)
    {
        IRasterLayer rasterLayer=new RasterLayerClass();
        IRaster raster=new Raster();
        raster=(IRaster)geoDataset;
        rasterLayer.CreateFromRaster(raster);
        rasterLayer.Name=interType;

        m_mapControl.AddLayer((ILayer)rasterLayer,0);
        m_mapControl.ActiveView.Refresh();
    }
```

23.4 功能调用

在【Spatial Analysis】页上添加【DEM Analyst】按钮。建立 Click 响应函数，代码如下：

```
private void btnEMAnalyst_Click(object sender,EventArgs e)
{
    DemAnalysisFrm frm=new DemAnalysisFrm(m_sceneControl);
    frm.ShowDialog();
}
```

23.5 编译测试

按下 F5 键，编译运行程序。

运行程序，点击按钮“DEM Analyst”，弹出分析窗口，添加输入图层，并设置输出文件路径和文件名。

附录 1：创建 SQLExpress 地理数据库

1. 安装 SQL Server Express 2016

安装过程和 SQL Server 2016 类似，注意选择 Windows 身份验证模式。

2. 安装 Database Server

ArcGIS 专门为 SQL Express 配套了数据库服务程序，在 ArcGIS 安装盘中可找到，如附图 1-1 所示。

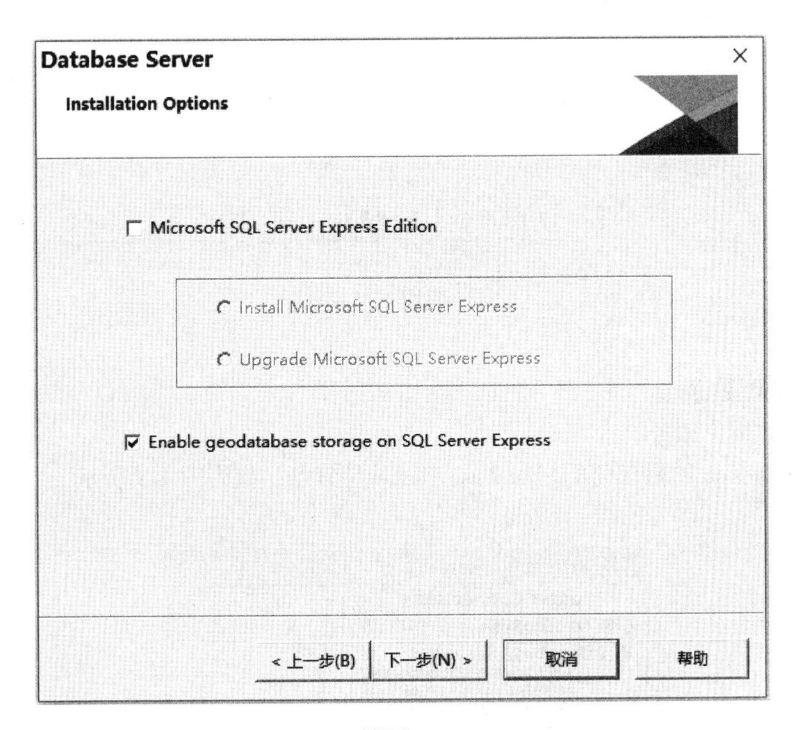

附图 1-1

勾选“Enable geodatabase storage on SQL Server Express”，点击【下一步】，如附图 1-2 所示。

在 SQL server 实例名下拉框中选“WIN-KG9LKA8CBST \ SQLEXPRESS”，Windows 登录名下拉框中选择“WIN-KG9LKA8CBST \ Administrator”，然后点击【下一步】即可完成安装。

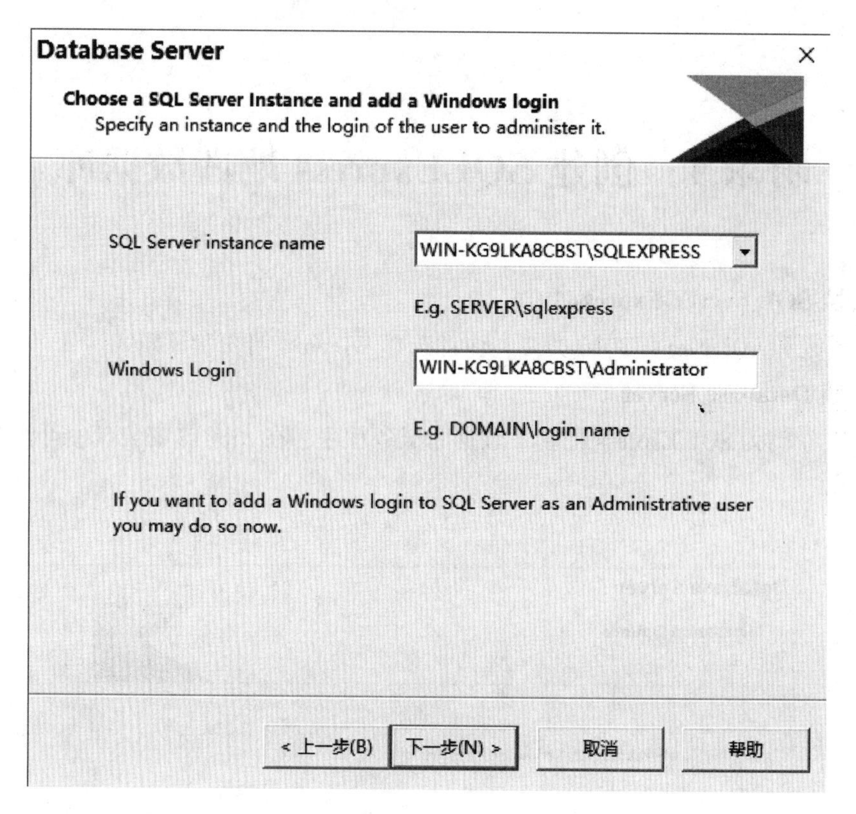

附图 1-2

3. 创建地理数据库

(1)创建数据库服务

找到 ArcCatalog 的目录树的【Database Dervers】节点，如附图 1-3 所示。

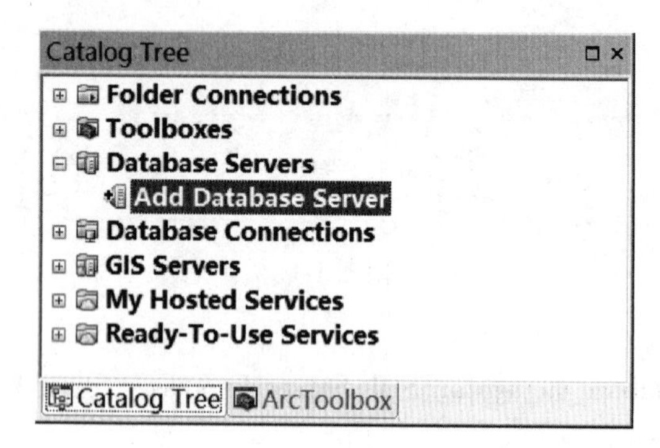

附图 1-3

双击【Add Database Server】，弹出窗口，如附图 1-4 所示。

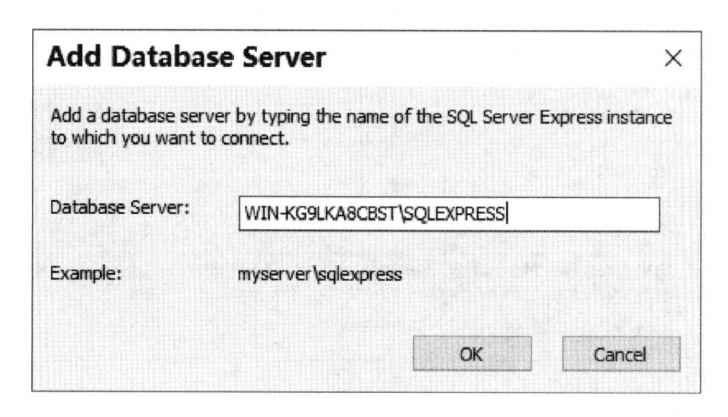

附图 1-4

设置 Database Server 为“WIN-KG9LKA8CBST\SQLEXPRESS”，点击【OK】，如附图 1-5 所示，出现数据库服务：WIN-KG9LKA8CBST_SQLEXPRESS_GDS。

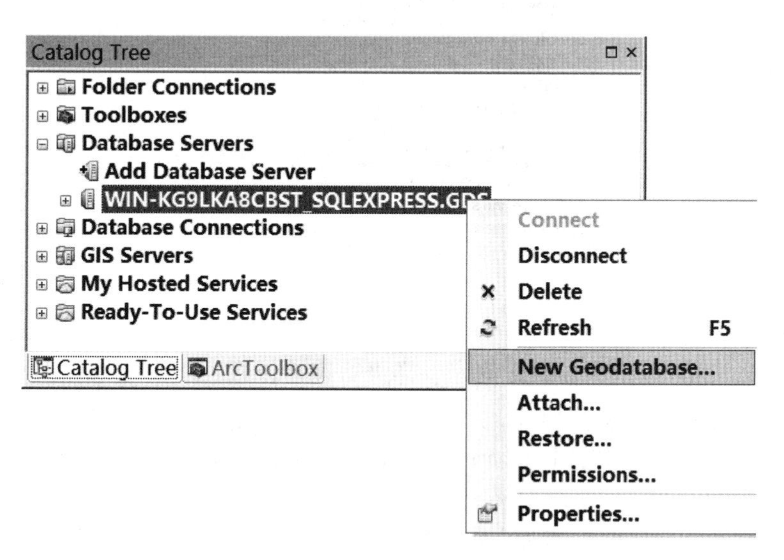

附图 1-5

(2)创建地理数据库

右键点击数据库服务“WIN-KG9LKA8CBST_SQLEXPRESS_GDS”，出现浮动菜单(附图 1-5)，点击“New Geodatabase”菜单项，出现地理数据库创建对话框，如附图 1-6 所示。

输入数据库名称(本例为 gdb_express)，选择数据库存放位置(本例为默认值)，数据库初始容量(本例为默认值：100Mb)，点击【OK】，如附图 1-7 所示，显示新的地理数据库在 SQL Express 中创建完成。

4. 数据库连接

地理数据库创建完成后，可以在 EXPRESS 地理数据库上进行数据库操作(右键点击

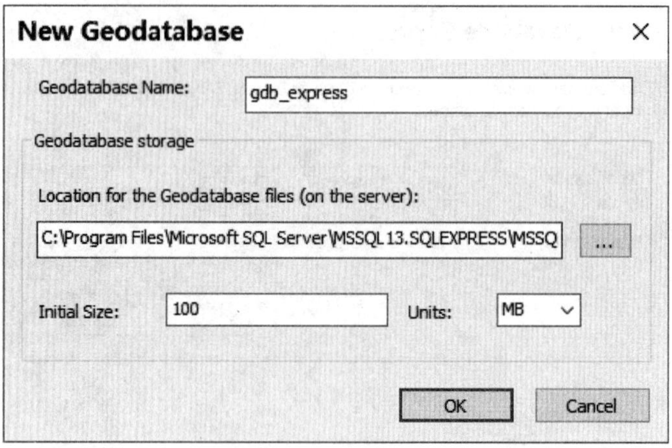

New Geodatabase
Geodatabase Name:
gdb_express
Geodatabase storage
Location for the Geodatabase files (on the server):
C:\Program Files\Microsoft SQL Server\MSSQL13.SQLEXPRESS\MSSQ
Initial Size:
100
Units:
MB
OK
Cancel

附图 1-6

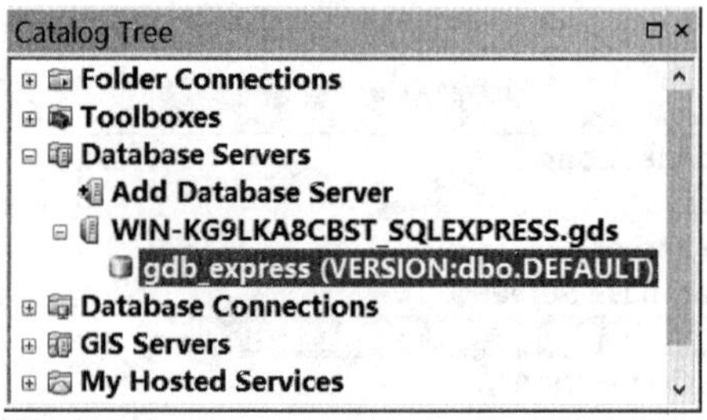

Catalog Tree
Folder Connections
Toolboxes
Database Servers
Add Database Server
WIN-KG9LKA8CBST_SQLEXPRESS.gds
gdb_express (VERSION:dbo.DEFAULT)
Database Connections
GIS Servers
My Hosted Services

附图 1-7

出现操作菜单)，也可以像企业级数据库一样进行连接设置，然后在链接节点上进行数据库操作。

找到 ArcCatalog 目录树的"Database Connections"节点，如附图 1-8 所示。

Catalog Tree
Folder Connections
Toolboxes
Database Servers
Add Database Server
Database Connections
Add Database Connection
GIS Servers
My Hosted Services
Ready-To-Use Services

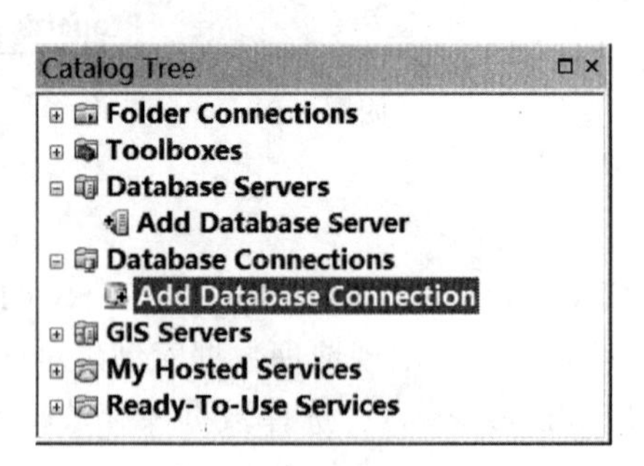

附图 1-8

双击【Add Database Connection】，弹出直连参数设置窗口，设置直连参数，如附图 1-9 所示。

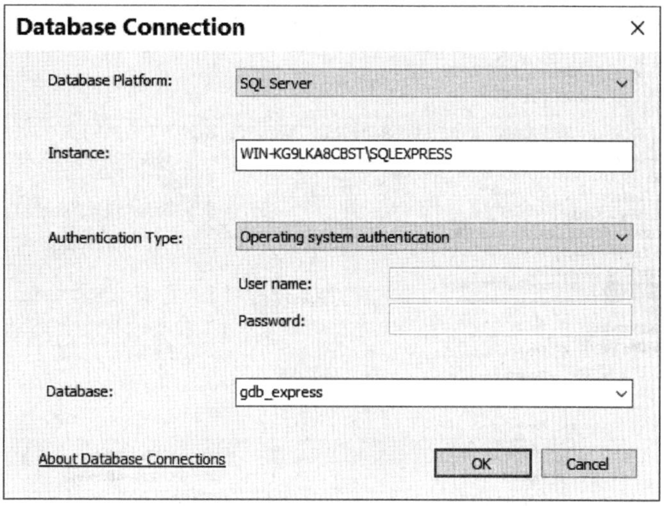

Database Connection
Database Platform: SQL Server
Instance: WIN-KG9LKA8CBST\SQLEXPRESS
Authentication Type: Operating system authentication
User name:
Password:
Database: gdb_express
About Database Connections
OK
Cancel

附图 1-9

注意：这里的 Instance 本例是 WIN-KG9LKA8CBST\SQLEXPRESS，然后选择要连接的地理数据库，选择操作系统验证(即 Windows 验证)及数据库名(本例是 gdb_express)。

点击【OK】，即可连接，如附图 1-10 所示。

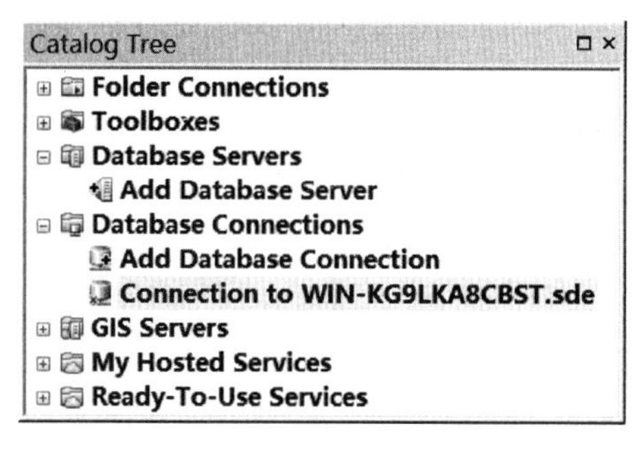

附图 1-10

5. 数据准备

(1)导入 shapefile 数据

在建立了空间数据库连接之后，我们就可以使用空间数据库进行工作。下面，我们把已有的 shapefile 数据通过 ArcSDE 导入到数据库中。

①右键单击“WIN-KG9LKA8CBST. sde”连接，选择 Import FeatureClass(Single)，如附图 1-11 所示。

②在弹出的对话框“FeatureClass to FeatureClass”中，点击 Input Features 右边的浏览按钮，打开浏览对话框，如附图 1-12 所示。

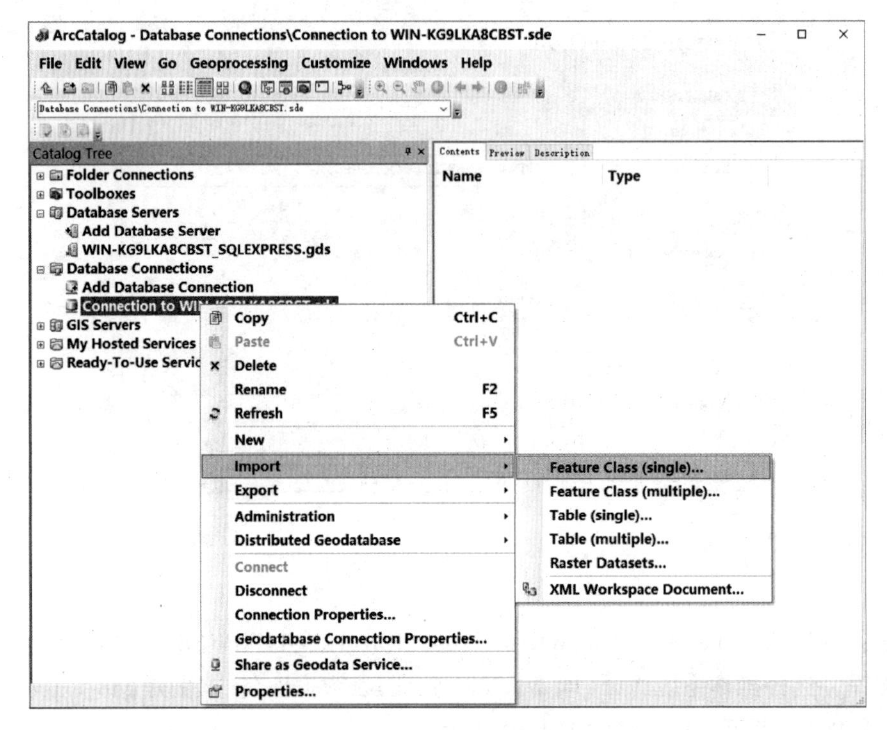
ArcCatalog - Database Connections\Connection to WIN-KG9LKA8CBST.sde
File Edit View Go Geoprocessing Customize Windows Help
Catalog Tree
Contents Preview Description
Name Type
Folder Connections
Toolboxes
Database Servers
Add Database Server
WIN-KG9LKA8CBST_SQLEXPRESS.gds
Database Connections
Add Database Connection
GIS Servers
My Hosted Services
Copy Ctrl+C
Paste Ctrl+V
Delete
Rename F2
Refresh F5
New
Import
Export
Administration
Distributed Geodatabase
Connect
Disconnect
Connection Properties...
Geodatabase Connection Properties...
Share as Geodata Service...
Properties...
Feature Class (single)...
Feature Class (multiple)...
Table (single)...
Table (multiple)...
Raster Datasets...
XML Workspace Document...

附图 1-11

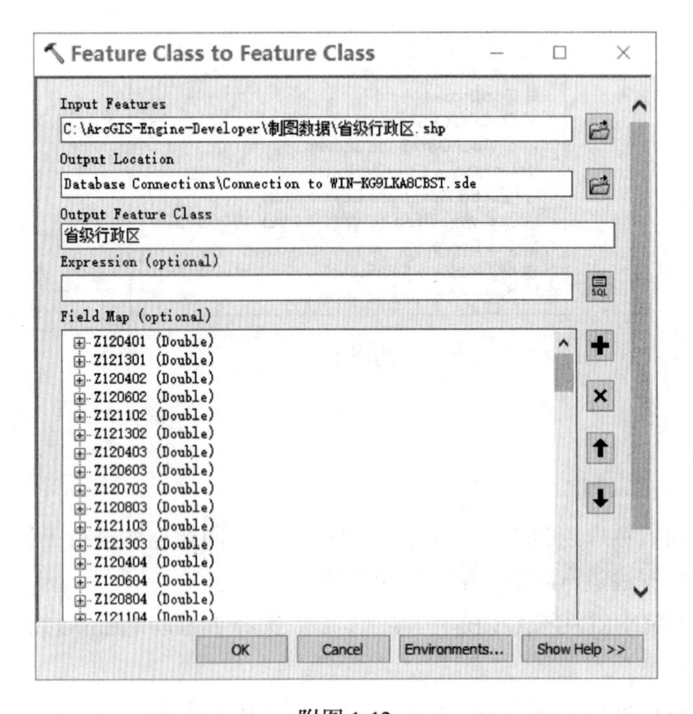
Feature Class to Feature Class
Input Features
C:\ArcGIS-Engine-Developer\制图数据\省级行政区.shp
Output Location
Database Connections\Connection to WIN-KG9LKA8CBST.sde
Output Feature Class
省级行政区
Expression (optional)
Field Map (optional)
Z120401 (Double)
Z121301 (Double)
Z120402 (Double)
Z120602 (Double)
Z121102 (Double)
Z121302 (Double)
Z120403 (Double)
Z120603 (Double)
Z120703 (Double)
Z120803 (Double)
Z121103 (Double)
Z121303 (Double)
Z120404 (Double)
Z120604 (Double)
Z120804 (Double)
OK
Cancel
Environments...
Show Help >>

附图 1-12

③浏览到需要导入的数据（本例中使用“C：\ArcGIS-Engine-Developer\制图数据\省级行政区.shp”数据），点击 ADD。

④在 Output FeatureClass 一栏中输入名称（输出要素类的名称），点击【OK】继续。

⑤在操作结束时，将出现提示对话框，点击“Close”继续。

⑥在“WIN-KG9LKA8CBST.sde”连接下可以看到导入进来的数据。至此，数据导入完成。

（注意：功能和个人数据导入功能完全相同。）

（2）导入栅格数据

ArcSDE 支持多种栅格数据格式，下面我们以 tif 数据为例来说明 ArcSDE 导入栅格数据的步骤。

①右键单击“WIN-KG9LKA8CBST.sde”连接，选择 Import Raster Datasets。

②在弹出的对话框中，点击 Input Features 右边的浏览按钮，打开浏览对话框。

③浏览到需要导入的数据（本例中使用“C：\GIS-Data\wsiearth.tif”数据），点击 ADD。

④在对话框中确认输入信息无误，点击【OK】继续。

⑤在导入结束时，将出现提示对话框，点击“Close”继续，如附图 1-13 所示。

附图 1-13

附录 2：ArcSDE 10. x 安装配置与连接

1. 概述

ArcSDE 10. x 的安装配置相较于 ArcSDE 10. 0 和之前版本，有了一些显著的变化，比如取消了 Post Install 向导，很多之前的管理操作改为使用地理处理工具来执行。本文以 ArcSDE 10. 1 为例介绍 ArcSDE 安装、配置和连接。安装环境如下：

①测试数据库：Microsoft SQL Server 2016 Enterprise Edition SP1；

②操作系统：Windows 7 SP1(64 位)，机器名叫 WIN-KG9LKA8CBST，注意系统防火墙需要关闭。

2. 安装

(1)SQL Server 2016 安装

这里不多讲 SQL Server 2016 的安装，只是说明一下其中需要注意的几个地方。

1)目录配置

要注意数据存储目录，默认的目录是 SQL Server 安装目录下的子目录，因此如果需要将数据存放到其他磁盘或路径，需要在这里制定数据根目录，如附图 2-1 所示。

附图 2-1

2) 实例配置

这里需要注意的是使用默认配置还是使用命名实例，如附图 2-2 所示，使用了默认实例。

附图 2-2

3) 身份验证

这里注意选择 Windows 身份验证，还是混合验证。选择第一种比较简单，选择第二种需要设置数据库用户名和密码，如附图 2-3 所示。

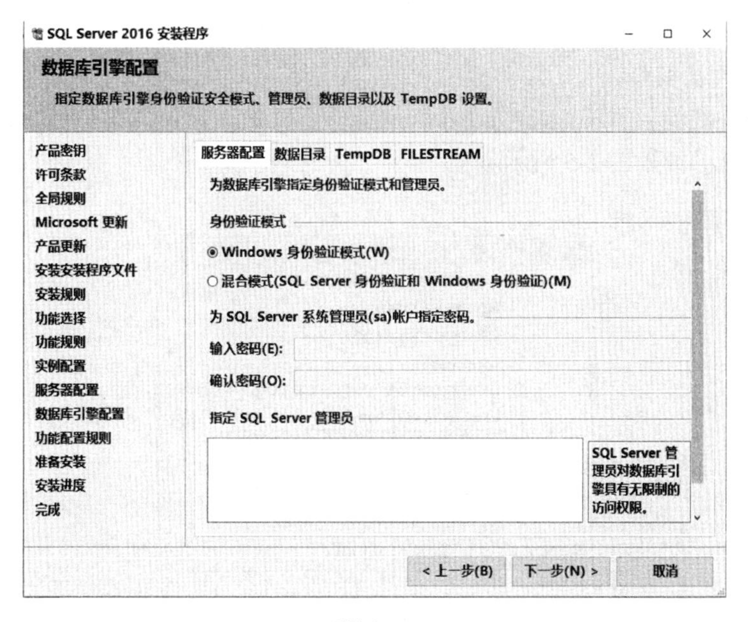

附图 2-3

(2)ArcSDE 10. x 安装

安装了 SQL Server 2016 之后，就可以安装 ArcSDE 了，如附图 2-4 所示，选择对应的安装项。

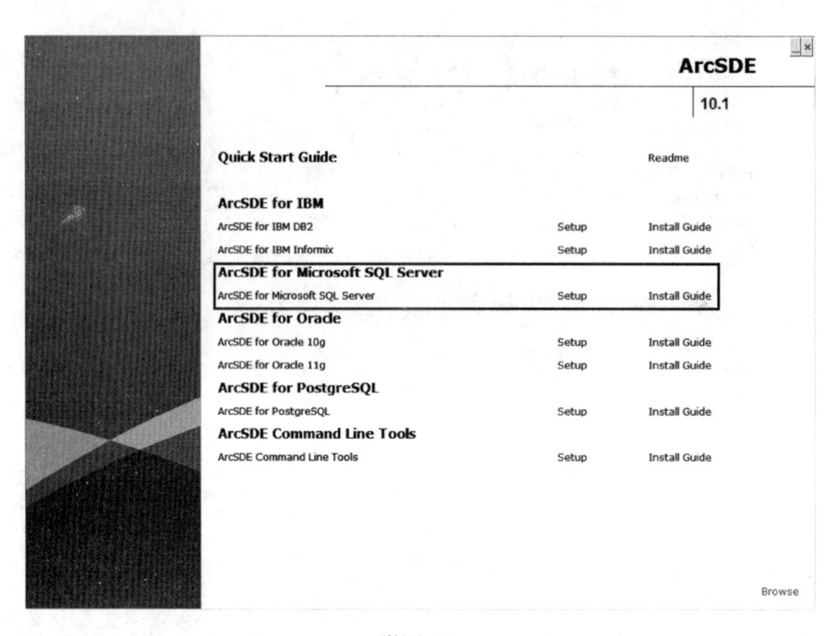

附图 2-4

ArcSDE 软件的安装没有什么特殊之处，只要一直 Next 就行了。安装完 ArcSDE 之后，没有像以前一样弹出 Post Install 向导，需要按照下面的步骤进行配置。进行配置的机器上需要已经安装 ArcGIS Desktop 10. x，以便使用地理处理工具。

3. 创建地理数据库

在工具箱中找到【Geodatabase Administration】工具集，其中包含进行地理数据库管理操作的若干工具，如附图 2-5 所示。

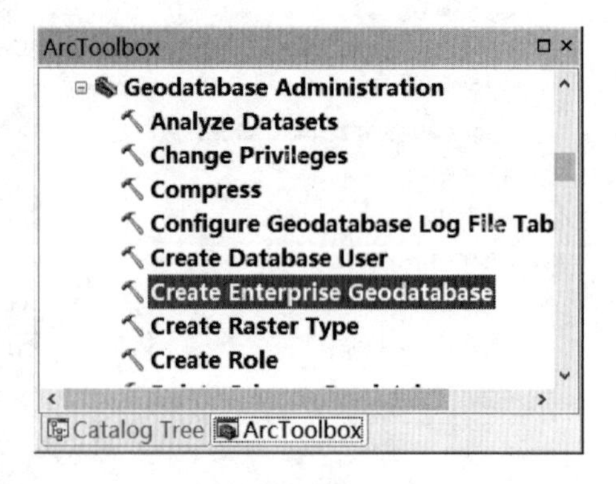

附图 2-5

双击打开【Create Enterprise Geodatabase】工具，输入参数，如附图 2-6 所示。

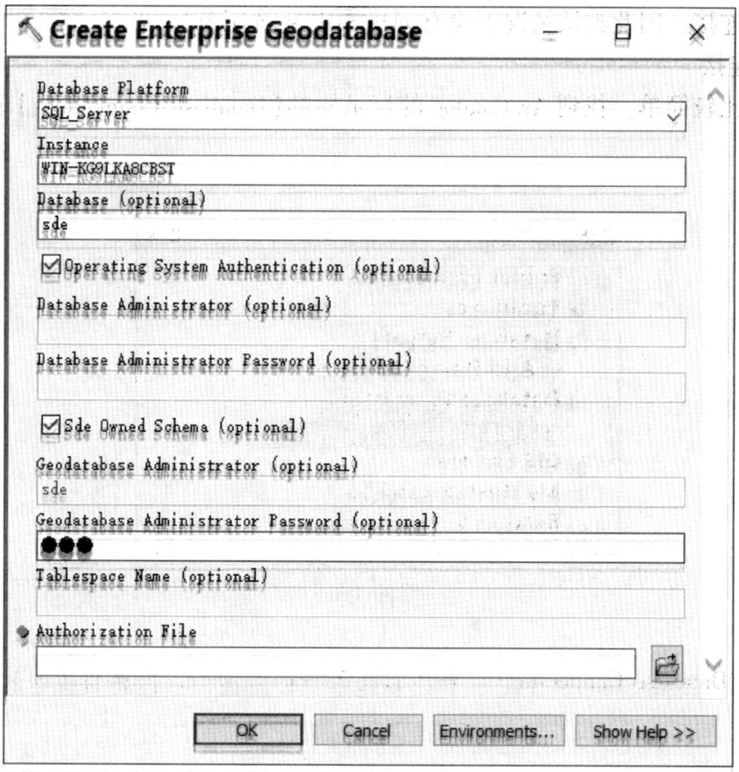

Create Enterprise Geodatabase
Database Platform
SQL_Server
Instance
WIN-KG9LKA8CBST
Database (optional)
sde
Operating System Authentication (optional)
Database Administrator (optional)
Database Administrator Password (optional)
Sde Owned Schema (optional)
Geodatabase Administrator (optional)
sde
Geodatabase Administrator Password (optional)
●●●
Tablespace Name (optional)
Authorization File
OK Cancel Environments... Show Help >>

附图 2-6

参数说明：

①Database Platform：数据库平台，本文选择 SQL_Server；

②Instance：SQL Server 实例名，本文是“WIN-KG9LKA8CBST”；

③Database：地理数据库名称，默认是 sde，也可以填其他名称；

④Authorization File：授权文件，即 .ecp 文件，一般和 ArcGIS Server 的授权文件是同一个。

以下两项可选，如果数据库验证选用 Windows 身份验证，不需要：

①Database Administrator(optional)：输入数据库管理员名；

②Database Administrator Password(optional)：输入数据库管理员密码。

以下两项可选：

①Geodatabase Administrator：输入地理数据库管理员名；

②Geodatabase Administrator Password：输入地理数据库管理员密码。

设置好参数后，点击【OK】开始创建地理数据库，此过程即相当于原来的 Post Install，将创建 SDE 系统表等。

4. 数据库连接

地理数据库已经创建成功，接下来就可以连接到地理数据库了。ArcGIS 10. x 中推荐使用直连方式连接，因此默认情况下没有创建 SDE 系统服务。

(1)直连连接

直连连接比较简单，找到 ArcCatalog 的目录树的【Database Connections】节点，如附图 2-7 所示。

Catalog Tree
Folder Connections
Toolboxes
Database Servers
Add Database Server
Database Connections
Add Database Connection
GIS Servers
My Hosted Services
Ready-To-Use Services

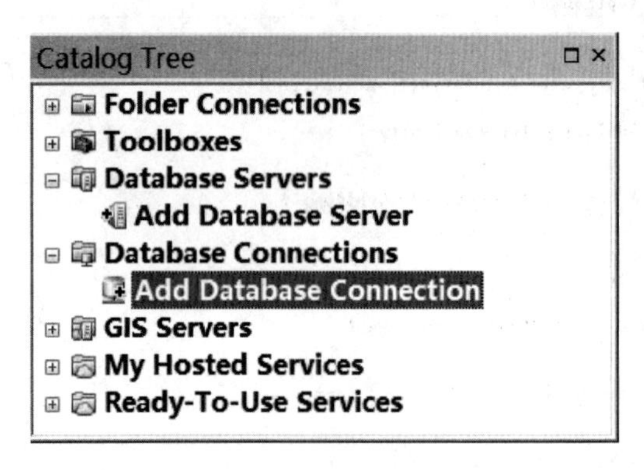

附图 2-7

双击"Add Database Connection"，弹出直连参数设置窗口，设置直连参数，如附图 2-8 所示。

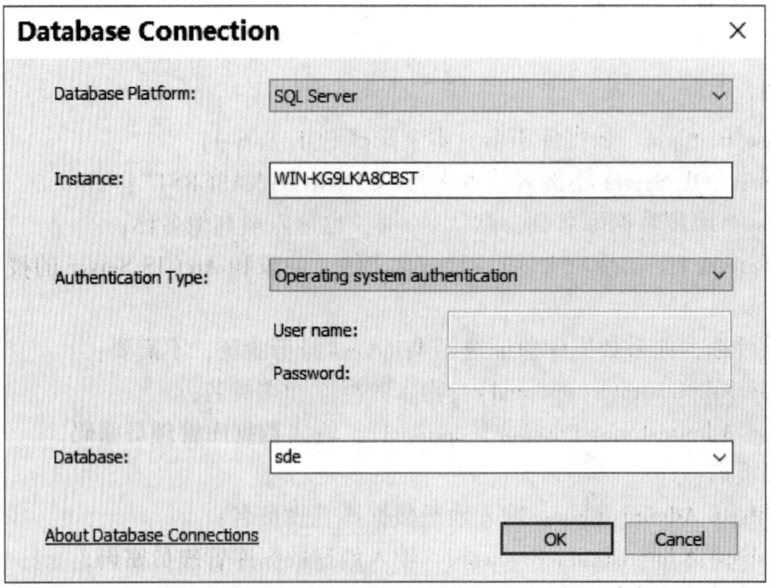

Database Connection
Database Platform:
SQL Server
Instance:
WIN-KG9LKA8CBST
Authentication Type:
Operating system authentication
User name:
Password:
Database:
sde
About Database Connections
OK
Cancel

附图 2-8

注意这里的 Instance，和创建地理数据库时的需要保持一致，然后选择要连接的地理数据库，输入用户密码，验证类型可选操作系统验证（即 Windows 验证）或数据库验证。

点击【OK】，即可连接，如附图 2-9 所示。

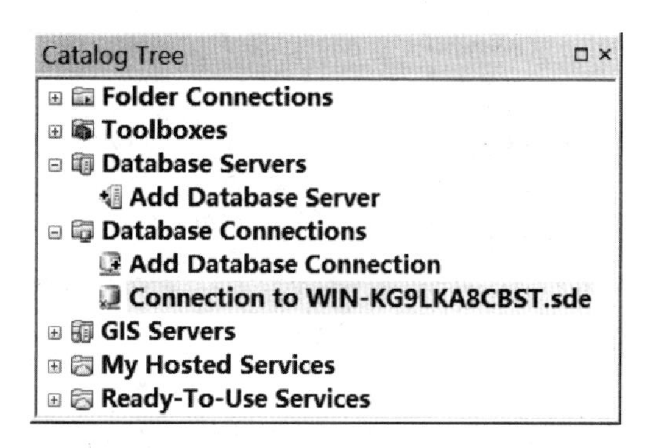

附图 2-9

（2）服务连接

除了推荐的直连方式，有时候我们还需要使用服务连接的方式，需要首先创建 ArcSDE 服务，然后在客户端以 . sde 连接文件的方式来连接地理数据库。

1）创建 ArcSDE 服务

创建 ArcSDE 服务需要经过三个主要步骤：手动修改服务文件、命令行安装服务、启动服务。

首先是手动修改 service. etc 文件，包括 ArcSDE 的 service 文件和 Windows 系统的 services 文件，如附图 2-10 和附图 2-11 所示。

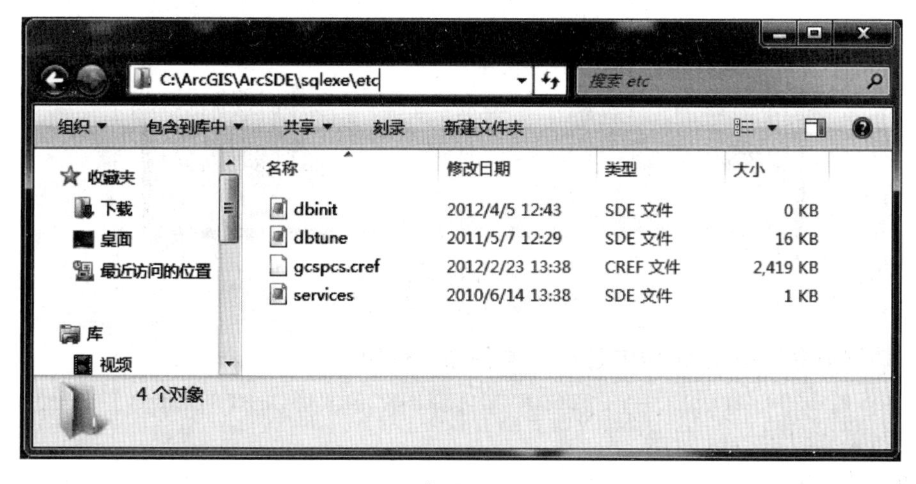

附图 2-10

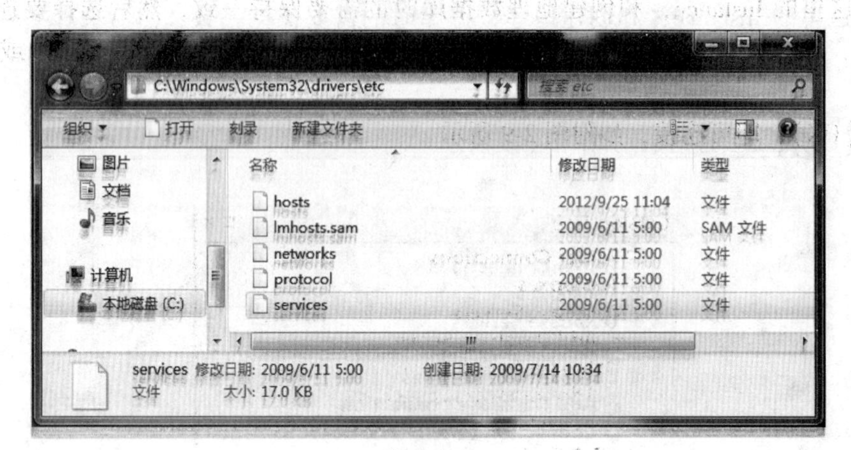

附图 2-11

在这两个文件中，分别添加“esri_sde 5151”并保存，如附图 2-12，附图 2-13 所示。

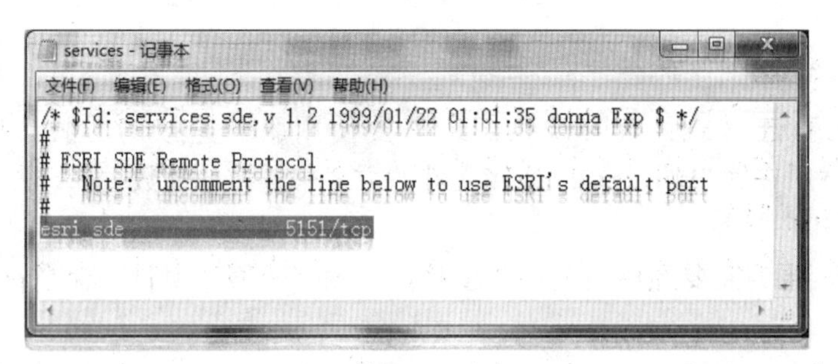

附图 2-12

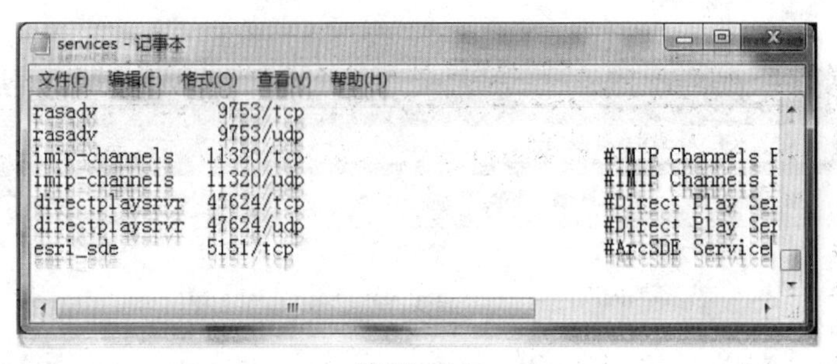

附图 2-13

然后使用命令行创建 ArcSDE 服务，本例命令行如下：

```
sdeservice -o create -d SQLSERVER,WIN-KG9LKA8CBST -psde -i esri_sde
```

执行结果如附图 2-14 所示。

创建成功后即可启动服务，命令行如下：

```
sdemon -o start -i esri_sde -p sde
```

附图 2-14

执行后将要求输入 sde 用户密码，然后即可启动，如附图 2-15 所示。

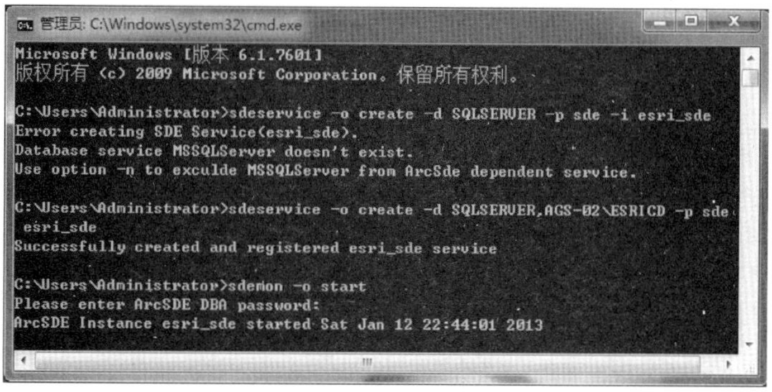

```
管理员: C:\Windows\system32\cmd.exe
Microsoft Windows [版本 6.1.7601]
版权所有 (c) 2009 Microsoft Corporation。保留所有权利。

C:\Users\Administrator>sdeservice -o create -d SQLSERVER -p sde -i esri_sde
Error creating SDE Service(esri_sde).
Database service MSSQLServer doesn't exist.
Use option -n to exculde MSSQLServer from ArcSde dependent service.

C:\Users\Administrator>sdeservice -o create -d SQLSERVER,AGS-02\ESRICD -p sde
esri_sde
Successfully created and registered esri_sde service

C:\Users\Administrator>sdemon -o start
Please enter ArcSDE DBA password:
ArcSDE Instance esri_sde started Sat Jan 12 22:44:01 2013
```

附图 2-15

SDE 服务已经启动，然后可以通过服务方式连接了。

2)创建服务连接文件

首先在【Geodatabase Administration】下【Workspace】工具集中找到【Create ArcSDE Connection File】工具，如附图 2-16 所示。

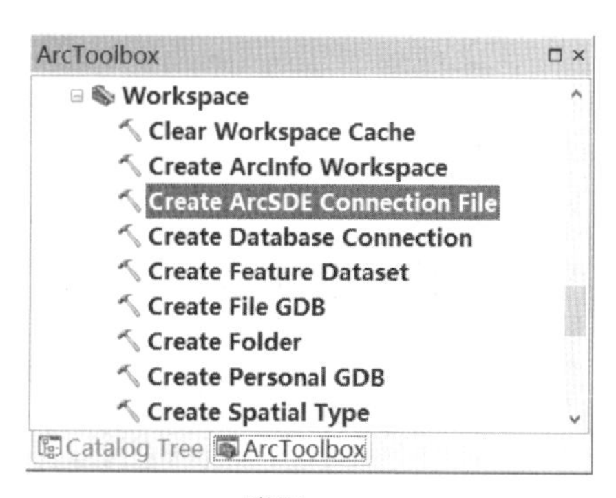

附图 2-16

打开工具，输入参数，如附图 2-17 所示。

Create ArcSDE Connection File

ArcSDE Connection File Location

C:\ArcGIS-Engine-Developer\制图数据\NewConnectiton

ArcSDE Connection File Name

esricd

Server

WIN-KG9LKA8CBST

Service

esri_sde

Database (optional)

sde

Database authentication (optional)

Username (optional)

sde

Password (optional)

●●●

Save username and password (optional)

The following transactional version will be used (optional)

Save the transactional version name with the connection file (optional

OK　Cancel　Environments...　Show Help >>

附图 2-17

这里需要设置 SDE 服务所在的机器名或 IP(本例是 WIN-KG9LKA8CBST)，SDE 服务名称(本例是 esri_ sde 或 5151)，此外输入数据库名称(本例是 sde)以及用户名密码。

点击【OK】即可完成 . sde 服务连接文件的创建，然后到指定目录下找到连接文件，双击即可连接到地理数据库，如附图 2-18 所示(上面直连测试中导入的数据清晰可见)。

附图 2-18

说明服务连接方式已经可以正常使用 ArcSDE 库了。

附录 3：LicenseInitializer 源代码

```
    internal sealed class LicenseInitializer
    {
        private IAoInitialize m_AoInit = new AoInitializeClass();

        #region Private members
        private const string MessageNoLicensesRequested = "Product: No
licenses were requested";
        private const string MessageProductAvailable = "Product: {0}:
Available";
        private const string MessageProductNotLicensed = "Product: {0}:
Not Licensed";
        private const string MessageExtensionAvailable = " Extension: {0}:
Available";
        private const string MessageExtensionNotLicensed = " Extension:
{0}:Not Licensed";
        private const string MessageExtensionFailed = " Extension: {0}:
Failed";
        private const string MessageExtensionUnavailable = " Extension:
{0}:Unavailable";

        private bool m_hasShutDown = false;
        private bool m_hasInitializeProduct = false;

        private List<int> m_requestedProducts;
        private List<esriLicenseExtensionCode> m_requestedExtensions;
        private Dictionary<esriLicenseProductCode, esriLicenseStatus>
m_productStatus = new Dictionary<esriLicenseProductCode, esriLicense-
Status>();
        private Dictionary<esriLicenseExtensionCode, esriLicenseStatus
> m_extensionStatus = new Dictionary<esriLicenseExtensionCode, esriLi-
censeStatus>();
```

```
        private bool m_productCheckOrdering = true; //default from low
to high
        #endregion

        public bool InitializeApplication(esriLicenseProductCode[]
productCodes,esriLicenseExtensionCode[] extensionLics)
        {
            //Cache product codes by enum int so can be sorted without
custom sorter
            m_requestedProducts = new List<int>();
            foreach (esriLicenseProductCode code in productCodes)
            {
            int requestCodeNum = Convert.ToInt32(code);
            if (!m_requestedProducts.Contains(requestCodeNum))
            {
                m_requestedProducts.Add(requestCodeNum);
            }
            }

            AddExtensions(extensionLics);
            return Initialize();
        }

        /// <summary>
        /// A summary of the status of product and extensions initializa-
tion.
        /// </summary>
        public string LicenseMessage()
        {
            string prodStatus = string.Empty;
            if (m_productStatus == null || m_productStatus.Count == 0)
            {
                prodStatus = MessageNoLicensesRequested + Environment.
NewLine;
            }
            else if (m_productStatus.ContainsValue(esriLicenseSta-
tus.esriLicenseAlreadyInitialized)
                    || m_productStatus.ContainsValue(esriLicense-
```

```
Status.esriLicenseCheckedOut))
              {
                  prodStatus = ReportInformation(m_AoInit as ILicenseIn-
formation,
                 m_AoInit.InitializedProduct(),
                 esriLicenseStatus.esriLicenseCheckedOut)+Environment.
NewLine;
              }
              else
              {
                 //Failed...
                  foreach (KeyValuePair<esriLicenseProductCode,esri-
LicenseStatus> item in m_productStatus)
                 {
                     prodStatus += ReportInformation(m_AoInit as ILi-
censeInformation,item.Key,item.Value)+Environment.NewLine;
                 }
              }

              string extStatus = string.Empty;
              foreach (KeyValuePair<esriLicenseExtensionCode,esriLi-
censeStatus> item in m_extensionStatus)
              {
                    string info = ReportInformation(m_AoInit as ILi-
censeInformation,item.Key,item.Value);
                 if (!string.IsNullOrEmpty(info))
                       extStatus += info+Environment.NewLine;
              }

              string status = prodStatus+extStatus;
              return status.Trim();
       }
       /// <summary>
       /// Shuts down AoInitialize object and check back in extensions
to ensure
       /// any ESRI libraries that have been used are unloaded in the
correct order.
```

```
        /// </summary>
        /// <remarks>Once Shutdown has been called,you cannot re-initialize the product license
        /// and should not make any ArcObjects call.
        /// </remarks>
        public void ShutdownApplication()
        {
                if (m_hasShutDown)
                    return;

                //Check back in extensions
                foreach (KeyValuePair<esriLicenseExtensionCode,esriLicenseStatus> item in m_extensionStatus)
                {
                        if (item.Value == esriLicenseStatus.esriLicenseCheckedOut)
                                m_AoInit.CheckInExtension(item.Key);
                }

                m_requestedProducts.Clear();
                m_requestedExtensions.Clear();
                m_extensionStatus.Clear();
                m_productStatus.Clear();
                m_AoInit.Shutdown();
                m_hasShutDown=true;
                //m_hasInitializeProduct=false;
        }

        /// <summary>
        /// Indicates if the extension is currently checked out.
        /// </summary>
        public bool IsExtensionCheckedOut(esriLicenseExtensionCode code)
        {
                return m_AoInit.IsExtensionCheckedOut(code);
        }

        /// <summary>
        /// Set the extension(s) to be checked out for your ArcObjects code.
```

```
        /// </summary>
        public bool AddExtensions(params esriLicenseExtensionCode[] re-
questCodes)
        {
                if (m_requestedExtensions==null)
                    m_requestedExtensions=new List<esriLicenseExtension-
Code>();
                foreach (esriLicenseExtensionCode code in requestCodes)
                {
                    if (!m_requestedExtensions.Contains(code))
                        m_requestedExtensions.Add(code);
                }

                if (m_hasInitializeProduct)
                        return CheckOutLicenses(this.InitializedProduct);

                return false;
        }

        /// <summary>
        /// Check in extension(s) when it is no longer needed.
        /// </summary>
        public void RemoveExtensions(params esriLicenseExtensionCode[]
requestCodes)
        {
                if (m_extensionStatus==null ||m_extensionStatus.Count==0)
                return;

                foreach (esriLicenseExtensionCode code in requestCodes)
                {
                    if (m_extensionStatus.ContainsKey(code))
                    {
                        if (m_AoInit.CheckInExtension(code)==
                                                                esriLicens-
eStatus.esriLicenseCheckedIn)
                        {
                                m_extensionStatus[code]=esriLicenseSta-
tus.esriLicenseCheckedIn;
```

```
                    }
                }
            }
        }

        /// <summary>
        /// Get/Set the ordering of product code checking. If true,check
from lowest to
        /// highest license. True by default.
        /// </summary>
        public bool InitializeLowerProductFirst
        {
            get
            {
                return m_productCheckOrdering;
            }
            set
            {
                m_productCheckOrdering=value;
            }
        }

        /// <summary>
        /// Retrieves the product code initialized in the ArcObjects ap-
plication
        /// </summary>
        public esriLicenseProductCode InitializedProduct
        {
            get
            {
                try
                {
                    return m_AoInit.InitializedProduct();
                }
                catch
                {
                    return 0;
                }
```

```
            }
        }
        #region Helper methods
        private bool Initialize()
        {
                if (m_requestedProducts = = null || m_requestedProducts.
Count = =0)
                    return false;

            esriLicenseProductCode currentProduct = new esriLicensePro-
ductCode();
            bool productInitialized=false;

            //Try to initialize a product
            ILicenseInformation licInfo=(ILicenseInformation)m_AoInit;

            m_requestedProducts.Sort();
            if (!InitializeLowerProductFirst) //Request license from high-
est to lowest
                m_requestedProducts.Reverse();

            foreach (int prodNumber in m_requestedProducts)
            {
                esriLicenseProductCode prod=
              (esriLicenseProductCode) Enum.ToObject (typeof (esriLi-
censeProductCode),prodNumber);
                 esriLicenseStatus status =m_AoInit.IsProductCodeAvailable
(prod);
                if (status = =esriLicenseStatus.esriLicenseAvailable)
                {
                        status=m_AoInit.Initialize(prod);
                           if (status = = esriLicenseStatus.esriLicenseAl-
readyInitialized ||
                               status = = esriLicenseStatus.esriLicense-
CheckedOut)
                        {
                            productInitialized=true;
                            currentProduct=m_AoInit.InitializedProduct();
```

```
                    }
                }

                m_productStatus.Add(prod,status);

                if (productInitialized)
                    break;
            }

            m_hasInitializeProduct=productInitialized;
            m_requestedProducts.Clear();

            //No product is initialized after trying all requested licen-
ses,quit
            if (!productInitialized)
            {
            return false;
            }

            //Check out extension licenses
            return CheckOutLicenses(currentProduct);
    }

    private bool CheckOutLicenses(esriLicenseProductCode currentProd-
uct)
    {
            bool allSuccessful=true;
            //Request extensions
            if (m_requestedExtensions!=null && currentProduct!=0)
            {
                foreach (esriLicenseExtensionCode ext in m_requestedEx-
tensions)
                {
                    esriLicenseStatus licenseStatus=m_AoInit.IsExten-
sionCodeAvailable(currentProduct,ext);
                    if (licenseStatus==esriLicenseStatus.esriLicenseAvail-
able)
                    //skip unavailable extensions
```

```
                {
                    licenseStatus=m_AoInit.CheckOutExtension(ext);
                }
                allSuccessful=(allSuccessful && licenseStatus==
                                                    esriLicenseS-
tatus.esriLicenseCheckedOut);
                if (m_extensionStatus.ContainsKey(ext))
                m_extensionStatus[ext]=licenseStatus;
                else
                m_extensionStatus.Add(ext,licenseStatus);
            }

            m_requestedExtensions.Clear();
        }

        return allSuccessful;
    }

    private string ReportInformation(ILicenseInformation licInfo,esri-
LicenseProductCode code,esriLicenseStatus status)
    {
        string prodName=string.Empty;
        try
        {
            prodName=licInfo.GetLicenseProductName(code);
        }
        catch
        {
            prodName=code.ToString();
        }

        string statusInfo=string.Empty;

        switch (status)
        {
        case esriLicenseStatus.esriLicenseAlreadyInitialized:
        case esriLicenseStatus.esriLicenseCheckedOut:
            statusInfo=string.Format(MessageProductAvailable,prodName);
```

```
                break;
            default:
                statusInfo=string.Format(MessageProductNotLicensed,prod-
Name);
                break;
            }

            return statusInfo;
    }
    private string ReportInformation(ILicenseInformation licInfo,esri-
LicenseExtensionCode code,esriLicenseStatus status)
    {
            string extensionName=string.Empty;
            try
            {
                extensionName=licInfo.GetLicenseExtensionName(code);
            }
            catch
            {
                extensionName=code.ToString();
            }

            string statusInfo=string.Empty;

            switch (status)
            {
            case esriLicenseStatus.esriLicenseAlreadyInitialized:
            case esriLicenseStatus.esriLicenseCheckedOut:
                statusInfo = string.Format(MessageExtensionAvailable,
extensionName);
                break;
            case esriLicenseStatus.esriLicenseCheckedIn:
                break;
            case esriLicenseStatus.esriLicenseUnavailable:
                statusInfo=string.Format(MessageExtensionUnavailable,
extensionName);
                break;
            case esriLicenseStatus.esriLicenseFailure:
```

```
                statusInfo = string.Format(MessageExtensionFailed,extensi-
onName);
                break;
            default:
                statusInfo = string.Format(MessageExtensionNotLicensed,
extensionName);
                break;
            }

            return statusInfo;
    }
    #endregion
    }
```

参 考 文 献

1. ESRI 中国(北京)培训中心. ArcGIS 轻松入门教程——ArcGISEngine. 2008.

2. ESRI 中国(北京)有跟公司. ArcGIS10 产品白皮书, 2010.

3. ESRI. ArcGIS Engine Help For. NET(VS 2015), 2015.

4. 汤国安. ArcGIS 地理信息系统空间分析实验教程[M]. 北京: 科学出版社, 2006.

5. 张丰, 杜振洪, 刘仁义. GIS 程序设计教程——基于 ArcGIS Engine 的 C#开发实例[M]. 杭州: 浙江大学出版社, 2012.

6. 李崇贵, 陈峰, 丰德息, 等. ArcGIS Engine 组件式开发及应用[M]. 北京: 科学出版社, 2012.

7. 荆平. 基于 C#的地理信息系统设计与开发[M]. 北京: 清华大学出版社, 2013.

8. 丘洪钢, 张青莲, 熊友谊. ArcGIS Engine 地理信息系统开发——从入门到精通[M]. 北京: 人民邮电出版社, 2013.

9. 荆平. 基于 C#的地理信息系统设计开发案例教程[M]. 北京: 清华大学出版社, 2014.

10. 芮小平, 于雪涛. 基于 C#语言的 ArcGIS Engine 开发基础与技巧[M]. 北京: 电子工业出版社, 2015.

11. 牟乃夏, 王海银, 李丹, 等. ArcGIS Engine 地理信息系统开发教程——基于 C#. NET[M].北京: 测绘出版社, 2015.